An olo.éditions production
www.oloeditions.com

Design and selection
Nicolas Marçais
Philippe Marchand

Editing
Fanny Marchal

Graphics
Philippe Marchand

Texts
Chantal Allès
Anne Kerloc'h
Nicolas Marçais

Layout and visuals
Manon Bucciarelli

© 2010 olo.éditions

Published by Vivays Publishing Ltd
www.vivays-publishing.com

This edition © 2011 Vivays Publishing Ltd

ISBN 978-1-908126-04-7

English translation: Carol Gullidge in association with First Edition Translations Ltd, Cambridge, UK

Edited by Andrew Whittaker

All rights reserved.
No part of this book may be reproduced in any form or by any electronic or mechanical means whatsoever without the written permission of the publisher.

Printed in January 2011

Printed in China by Imago

Foreword: not all of the products appearing in this book are inexhaustible, and there may well be some that are no longer available at the time you're looking for them. Don't panic: it's often possible to reserve them as soon as they are back in stock or, if you can't wait, to find similar products on other websites.

CRAZY
GIFTS

Vivays Publishing

Your beloved's birthday, the kids' Christmas presents, the in-laws' anniversary, an invitation to the Joneses, Jean-Pierre's send-off... Although there's no shortage of occasions for giving, it's unfortunately more often the ideas that are thin on the ground. How, and above all, where to find the ideal present, the one that's just perfect, which will delight the recipient and gratify the giver without putting too much of a strain on the budget? You no longer need to deploy all your imaginative powers, live in the capital, or be rolling in money. A computer, a few judicious clicks, and you're away! But you still need to know where to search and what to look for amongst the thousands of websites and the millions of products available on the Web.

To help you find that rare gem, this book contains gifts with a difference, which are available directly through the Internet. To help you lay your hands on the ideal gift without having to trudge around the shops, we have, wherever possible, indicated two online stores (one in the UK and the other in the US). With prices being so volatile, feel free to check against other websites to be sure of getting a good deal. A handy book that will rapidly help you find the very thing that will suit your budget, the type of recipient (party animal, green, etc.) and even the type of occasion (party, birthday, Valentine's Day...).

1
Kitchen & Dining

The Art of
Entertaining
Cocktails
Gourmet Delights
Cooking Utensils

2
Wellbeing

Body Treatments
Beauty
Exercise
Relaxation
Sex

3
Fashion

Bags
Jewellery
Clothing
Shoes
Accessories

4
Games & Toys

Thinking Games
Construction Toys
Outdoor Toys
Toys for Pets
Sports

5
Gadgets

Stationery
Office
Leisure
Accessories
Travel

6
Techno

Video
Audio
Computers
Multimedia
Electronics

7
Home

Entrance Hall
Bedroom
Lounge
Bathroom
Toilets

KITCHEN & DINING

10 THROWZINI'S KNIFE

KITCHEN | www.amazon.com | www.amazon.co.uk

Practical and fun, this handmade knife block features five knives sheathed in individual guards and a spinning wheel complete with nervous assistant. Finally, an incentive to put the knives away rather than simply leaving them on the draining board. Not one for the children!

FORKED UP

www.thout.ca   www.designpublic.com   KITCHEN

Give free rein to the psychotic knife-thrower that lurks somewhere deep (hopefully) in your subconscious with this wall-mounted cutlery storage system. Simply pick up knives, forks and spoons and fling at the magnetic panel, thus avoiding the boredom of filing your silverware in the traditional cutlery drawer. All the fun of the fair, and it looks great too.

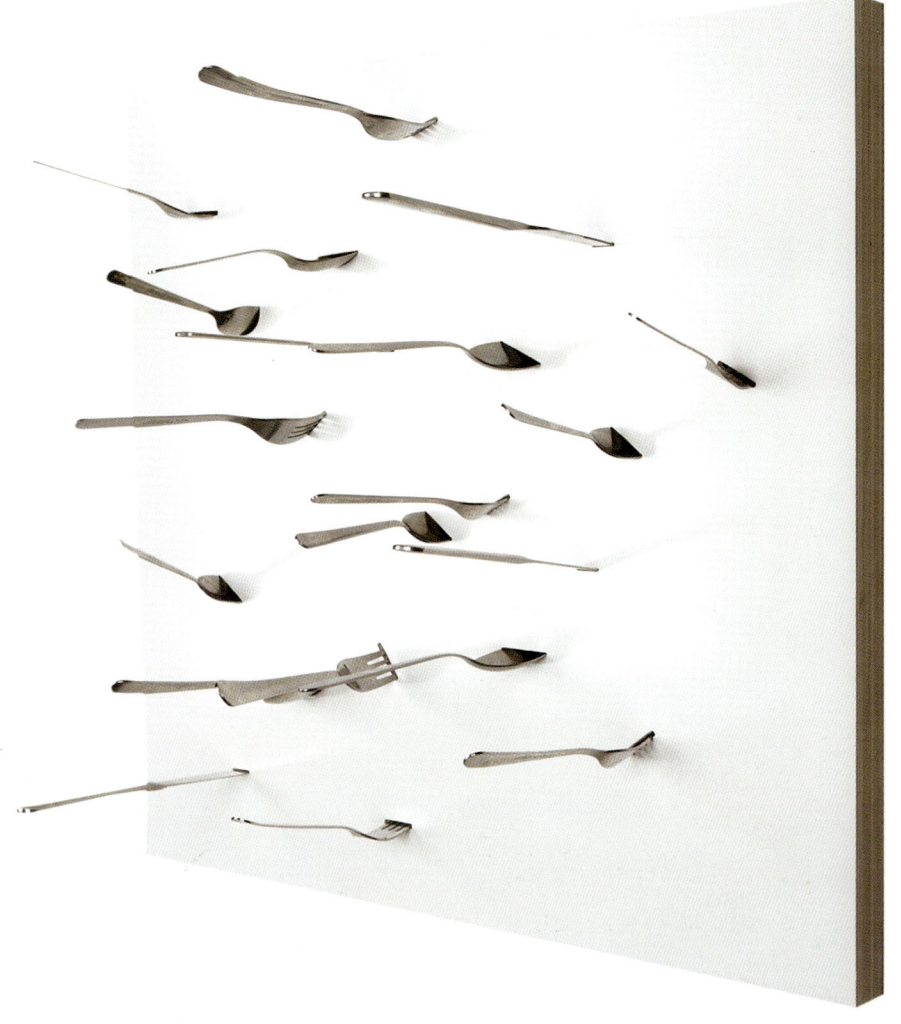

12 POUND PARTY PICK SET

KITCHEN | www.perpetualkid.com | www.find-me-a-gift.co.uk

There's something inherently sad about watching a grown man chasing an olive around a bowl with a slender wooden cocktail stick, jabbing away without success. Subtly put him out of his misery; present him with a three-inch 'cocktail nail'. The sticks are supplied with a wood-effect foam base, allowing you to keep him guessing whilst he works out their real purpose.

WENGER GIANT KNIFE

www.amazon.com   www.swiss-army-knife-wenger.co.uk

KITCHEN

A veritable workshop in miniature, the Giant Knife delivers 87 tools and 141 functions. A compass, magnifying glass, cigar cutter, tin opener, cuticle stick, drill, scissors, secateurs, pliers, screwdrivers, blades and yet more blades all feature in the armoury. Designed by Wenger, traditional maker of the Swiss army knife.

14 FOOT BOWL

KITCHEN | www.alexgarnett.com

Can I kick it? No, you can't – it's a ceramic bowl. The ideal gift for the sports mad nephew or the gourmet cook; or perhaps simply to cheer a deflated friend. Also available in a Basket Bowl version. Watch this space for the cricket bowl, the tennis bowl, the belle of the bowl, etc., etc.

FOOD FACE KID'S DINNER PLATE

www.perpetualkid.com   www.gadgets.co.uk   KITCHEN

15

We all know that kids love playing with their food, so why not channel the mealtime disobedience into something creative? Mashed potato becomes hair, beans become eyebrows, the gravy a five o'clock shadow — Picasso would have been proud. Who knows, the Food Face Dinner Plate might even persuade the little ones to eat their broccoli.

16 STRAWZ

KITCHEN | www.thinkgeek.com | www.firebox.com

The liquid journey from glass to mouth takes an adventurous turn with this, the straw to end all straws. With 20 rubber connectors and 16 translucent tubes, the DIY kit allows you to indulge that untapped creative bent, or simply offers an interesting diversion should conversation run dry. It sucks big time... but in a good way, of course.

MODERN-TWIST PLACEMATS

17

www.modern-twist.com   www.designpublic.com   **KITCHEN**

If you can't get the little darlings to focus on dinner itself, at least get them to focus on the venue for dinner, the table. Modern-Twist placemats can be coloured in using marker pens and then wiped clean (of both doodles and noodles) ready for the next mealtime. Never again will you have to roam the house shouting 'Dinner's ready'.

18 POP ART TOASTER

KITCHEN | www.poparttoasterstore.com | www.amazon.com

All other toasters are toast! The Pop Art Toaster gives your toast a personality, branding a design onto the bread. Four stencil plates come with each toaster, selected from a range that includes heart, flowers, smiley, snowflake and chick. You can even write a love letter through the medium of toast.

IIAMO SELF-HEATING BABY BOTTLE

www.iiamo.com

KITCHEN

The self-heating baby bottle is a triumph of practicality and sleek design. Created by leading designer Karim Rashid, the eco-friendly cartridge heats milk in less than five minutes without using any electricity, operating instead with an ingenious shaken trigger mechanism. It's also made without using any toxic materials. No wonder the Iiamo bottle has won two prizes for innovation.

20 GLOBAL WARMING MUG

KITCHEN | www.uncommongoods.com | www.firebox.com

One cup of hot Armageddon please. Fill the Global Warming Mug with a hot drink and watch as the world's shorelines gradually disappear below the waves. Goodbye Florida. So long Bangladesh. Set to simulate a sea level rise of a hundred metres, the natural disaster recedes with the heat of the drink, cooling to restore the coastlines we know and love.

ON/OFF MUG

21

www.amazon.com www.iwantoneofthose.com

KITCHEN

Is there anything worse? Barely awake, groping around in the morning for that much needed coffee cup, raising it to the lips and finding, arrgh, it's gone cold. Finally, salvation is here in the shape of an intelligent mug. Fill it with hot tea or coffee and the 'on' sign gradually appears, fading as the cup cools to reveal the 'off' sign. Think: white=hot; black=cold.

22	**SPLAT STAN COASTER**	
KITCHEN	www.thinkgeek.com	www.gadgets.co.uk

Dear departed Stan – literally 'mugged' in his prime – how we loved him. But how pleasing too that something useful could emerge from such a sad demise; his poor, misshapen midriff perfectly shaped to hold a nice warm cuppa safe and sound on your desktop. Surely, it's what Stan would have wanted.

FISTICUP

23

www.perpetualkid.com www.amazon.co.uk

KITCHEN

Is it me or is everyone in the office suddenly taking me more seriously? Nothing draws the respect of your colleagues quite like a knuckle-duster. Just don't get carried away; remember, it's only a mug. If nothing more, it will at least give some added punch to that early morning coffee.

24 — BBQ BRANDING IRON

KITCHEN | www.amazon.com | www.firebox.com

If you've got something to say, why be boring — say it with branded meat. A love note on a burger; your name on a steak — the BBQ Branding Iron lets you personalise your barbecues, using as many as 80 interchangeable letters to get your message across. Simply heat the iron on the coals and apply to the food.

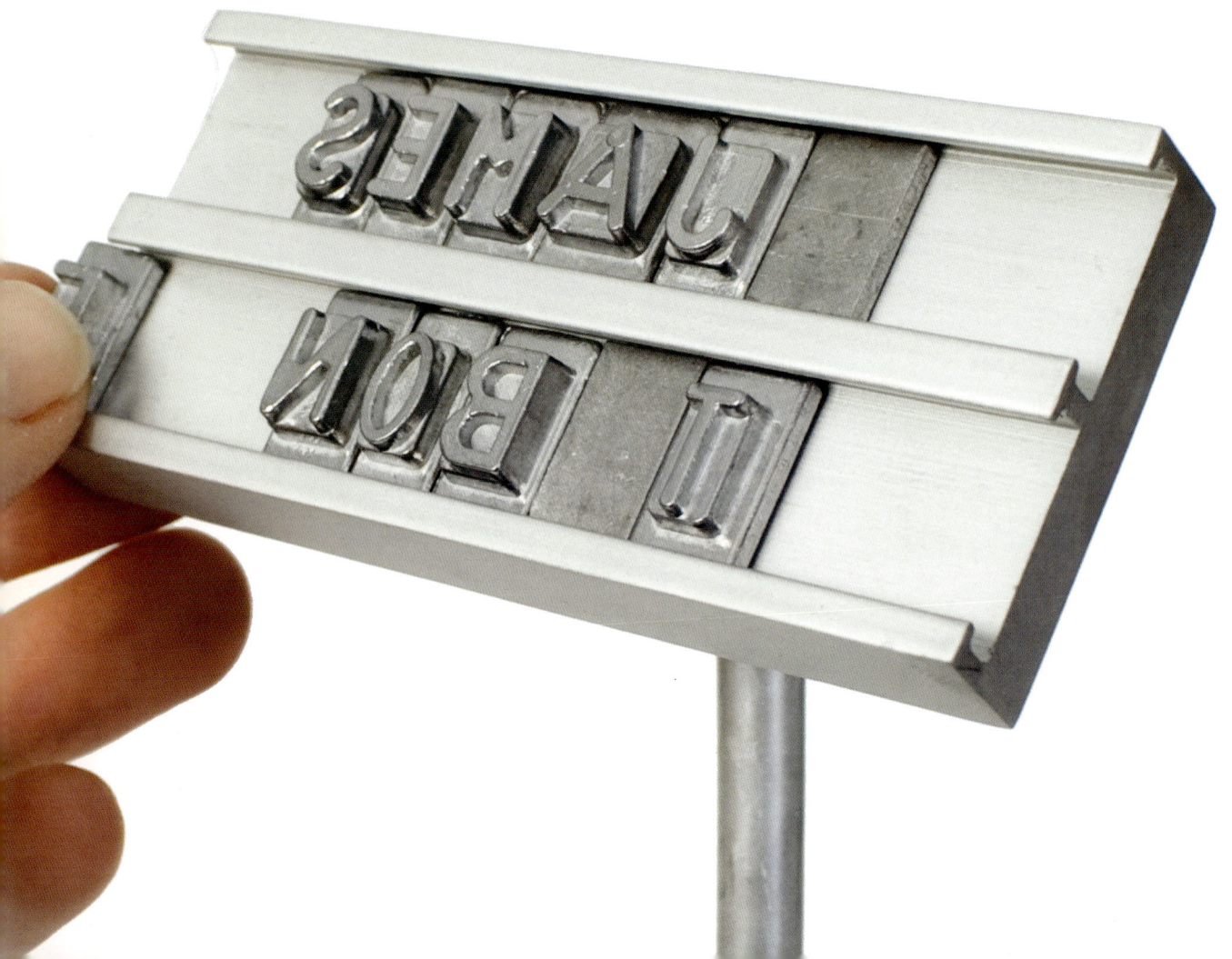

THE GODFATHER SPAGHETTI MD

www.david-louis.com

KITCHEN

The Godfather may look like an exhibit from a court case, but in truth it's the handiest new addition to your kitchen. It's a bilingual (English and Italian) spaghetti measurer, each finger hole sized to suit the pasta recipient, from 'child' to 'triplets' – say it in Italian for the full Godfather effect (ideally with cotton balls stuffed in the cheeks).

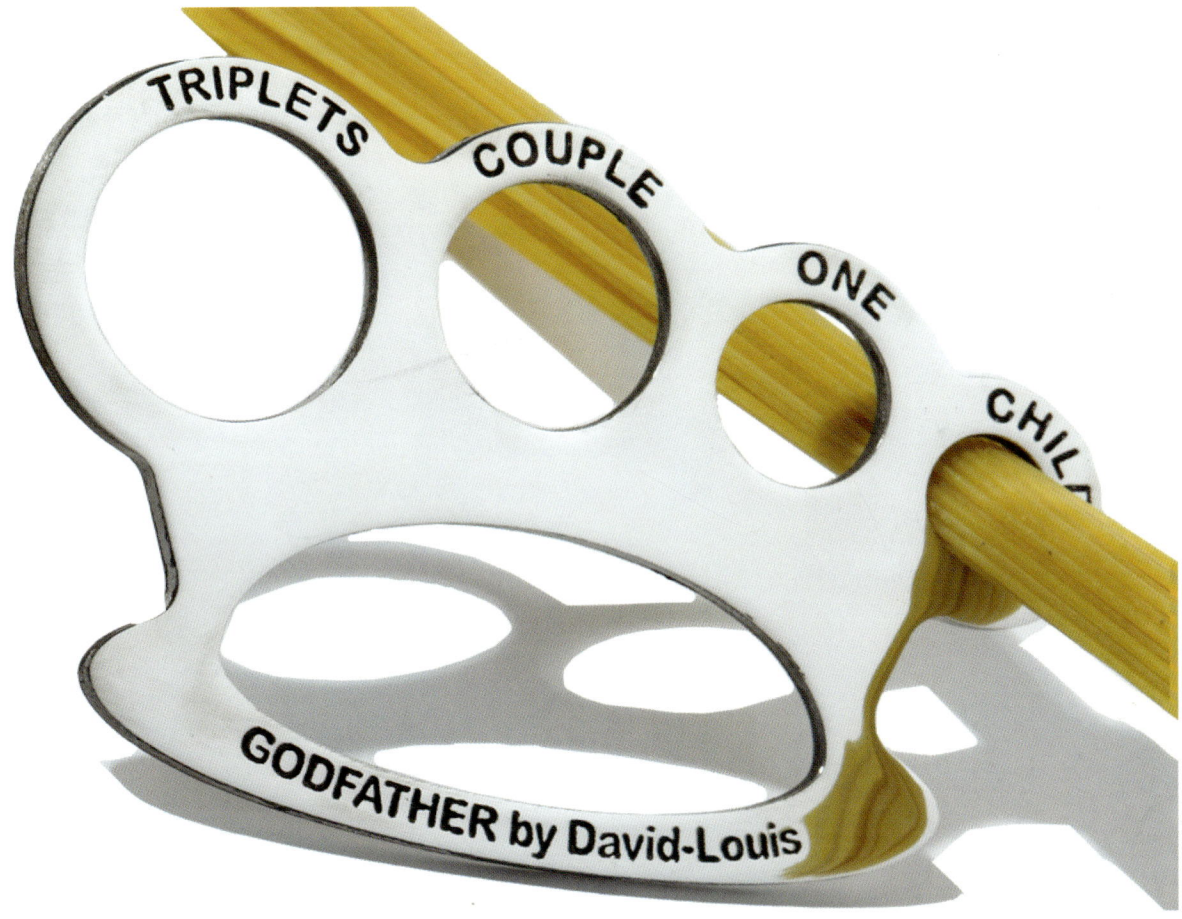

INSECT CANDIES

KITCHEN

www.hotlix.com

www.edible.com

"Look darling, canapés. Hmm, crunchy, almost earthy on the palate. What is it?" "Caterpillar." OK, it's probably best to warn recipients when you're offering the insect candies around at your next dinner party – the key thing is to remind them that it's all just protein. Take your pick from crickets, grasshoppers, ants, scorpions, caterpillars…

ANTI-THEFT LUNCH BAGS

www.thinkofthe.com   www.perpetualkid.com   **KITCHEN**

Sandwich envy: bane of the 21st century schoolyard, and perhaps even the office canteen. You can feel their eyes on your sarnie, mentally devouring your lovingly crafted lunch. What to do? Camouflage the sandwich with a small pouch of mouldy genius. That's fake mould of course. You need never padlock your lunchbox again.

28 LE WHIF

KITCHEN   www.lewhif.com

Who can deny the irresistible allure of chocolate? Even the strongest of us have given in to the craving at some time or other. It's sheer hell for anyone counting their calories. So, all hail David Edwards, the man who found the solution to our frustration in the shape of Le Whif, a chocolate inhaler. The organic chocolate powder relieves the craving with just one whiff.

'TEA PARTY' TEA BAGS

www.donkey-products.com www.pedlars.co.uk

KITCHEN

Do you dream of taking tea with the Queen? I'm sure she's been meaning to ask, but you know how busy her schedule is. What about the US President? Or the German Chancellor? Still no invite? Fear not, for these 'Tea Party' Tea Bags allow you to 'take tea' with the star of your choice, from footballers to striptease dancers, royals to rockers, Elvis Presley to Charlie Chaplin.

30 EVA SOLO TEA MAKER

KITCHEN www.amazon.com www.amazon.co.uk

It's easy to see why the Eva Solo Tea Maker won the international design prize at the Red Dot Award in 2005. The aesthetic appeal melds with impressive technology: for reduced tannin levels put the leaves directly in the teapot where they are held by a filter, or for a stronger drink insert the leaves in the filter and activate the piston. A neoprene jacket keeps the tea hot.

TERRORIST TEA POT

31

www.gnr8.biz   www.suck.uk.com   KITCHEN

A dissident teapot that simply refuses to work within the established teatime framework, but which does, thankfully, seem happy enough to brew a perfect cuppa; five or six cuppas in fact, for our balaclava-wearing malcontent is surprisingly roomy. Teatime has never been so radical.

32 CHAIN WINE BOTTLE HOLDER

KITCHEN | www.amazon.com | www.iwantoneofthose.com

It looks like a simple chain, yet when you slide a wine bottle – full or empty – into it, somehow it holds the bottle in place. As if by magic, the wine appears to be levitating. And it's nothing to do with the effects of alcohol, just the laws of gravity. Invite a science teacher round to dinner for a full explanation.

CALF & HALF CREAMER

www.perpetualkid.com   www.amazon.co.uk   KITCHEN

Where does milk come from? The supermarket. Yes, but before that? Err? Here's a jug that provides a daily reminder at the breakfast table, lest you forget the exact origins of that white stuff you're pouring on your cereal. At first sight it looks like an ordinary jug. However, pour in the milk and the unmistakable shape of the cow's udder appears. A great one for amusing the kids.

34 EDGE BROWNIE PAN

KITCHEN | www.bakersedge.com | www.firebox.com

Don't underestimate the science that goes into cooking a good brownie. How is it done? How do you get the perfect, slightly crispy outer layer combined with a soft, gooey centre? This zigzag baking tin makes it a whole lot easier, evenly dispersing the heat during cooking to achieve a runny centre and perfectly caramelised edges. Good work brownie boffins.

GUN EGG FRYER MOULD

www.amazon.com www.gidgit.co.uk KITCHEN

The breakfast routine can be, well, a bit routine. Every day, the same old story: kids squabbling, porridge bubbling. And look at those fried eggs; they're so egg shaped. Why can't eggs just be a bit more original? Give them a helping hand with this gun-shaped fried egg mould. Go on, start the day with a bang.

WELLBEING

2

38 SHOWER SHOCK CAFFEINATED SOAP

WELLBEING — www.thinkgeek.com

Shower shock, the first caffeinated soap, is the hygienic equivalent of drinking two cups of black coffee. The bar of soap releases 200 milligrams of caffeine per average soaping, utilising the skin's natural permeability to leave you feeling refreshed and invigorated. Resist the temptation to add milk, sugar or a wee dram.

BEER SOAP

www.thebeersoapcompany.com | www.etsy.com | **WELLBEING**

Beer soap isn't a gimmick. Granted, it is composed largely of beer, but its raison d'être is as a tonic for the skin, hair and nails. The soap can also be of benefit to sufferers of eczema and acne. And don't worry; you won't step out of the shower smelling of ale — the soap doesn't leave an odour on the body. Available in a range of flavours, including Guinness, Heineken and Corona.

40 GUN COMB

WELLBEING · www.25togo.com · www.pa-design.com

Is that a gun in your pocket or are you just pleased to see me? No, and no. It's actually a comb. The days when the girls were impressed by a traditional comb whipped from the pocket and combed through a luxuriant quiff are long gone. In the 21st century you need something extra — and that's what the gun comb gives.

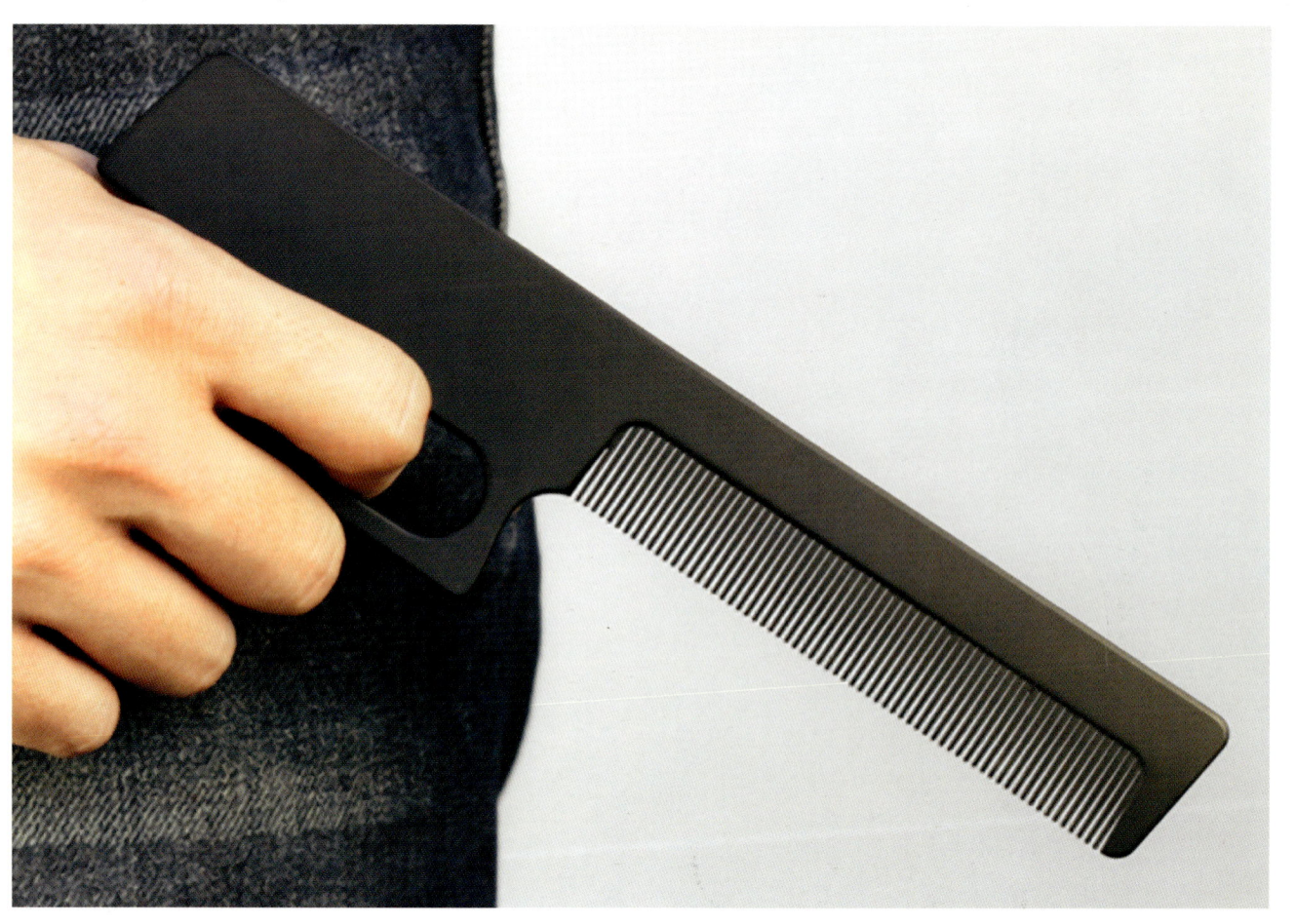

REALISTIC XBOX REPLICA SOAP

www.thinkgeek.com www.etsy.com **WELLBEING**

Ah, the pain of separation. No gamer likes to spend time apart from the dearly beloved joystick. Even taking a shower can be an ordeal, the unwelcome break from the gaming screen akin to water torture. The answer? A placebo. Or, to be more precise, a bar of realistic Xbox Replica Soap. Surely the easiest way to convince a game-obsessed youngster to take a shower.

42 GELICITY SPA JELLY BATH

WELLBEING — www.gelicity.com

Get the luxuriant spa experience in your own bathroom with this amazing blue gel. Infused with essences of bergamot, jasmine and lavender, Gelicity transforms from a simple powder into a thick creamy gel on contact with water, turning a normal bath into something distinctly heavenly. Helps to soften hard skin and clean deep into the skin lines.

CAVIAR CREAM

43

www.todoportv.com   www.orangecare.tv   **WELLBEING**

Renowned for centuries as the height of gastronomic luxury, caviar is also fast becoming the new beauty must-have for smoother, softer skin. Anti-wrinkle, firming and cell renewing, the caviar cream imbues the skin with essential fatty acids, vitamins A, B and D, and trace elements. The cream can be applied day or night.

44 DESIGN YOUR OWN FRAGRANCE KIT

WELLBEING | www.presentsandgiftideas.com | www.firebox.com

All the big celebs are designing their own perfumes these days, so why shouldn't everyone else do the same? This fragrance kit allows you to get in on the act. Each kit contains six bottles of fragrance (carefully selected to complement one another), which can be blended and engineered to create a new, unique scent.

SCENTS OF TIME

www.scentsoftime.co.uk

WELLBEING

Nothing is more evocative of time and place than scent, stirring up those deeply held memories of people and things. But what about the scents of the past? What were perfumes like in the times of Cleopatra, the Mayans, or Tutankhamen? Piecing together the aged fragments of evidence, David Pybus has recreated the fragrances that were fashionable in those ancient civilisations.

48 KINESIO TEX TAPE

WELLBEING   www.amazon.com   www.amazon.co.uk

Made from porous cotton that allows the skin to breathe, Kinesio Tex Tapes can relieve the aches and pains brought on by prolonged exertion without hampering movement. The tapes can also support muscles in movement, acting as a preventative to injury as well as helping to improve the awareness of posture and alignment.

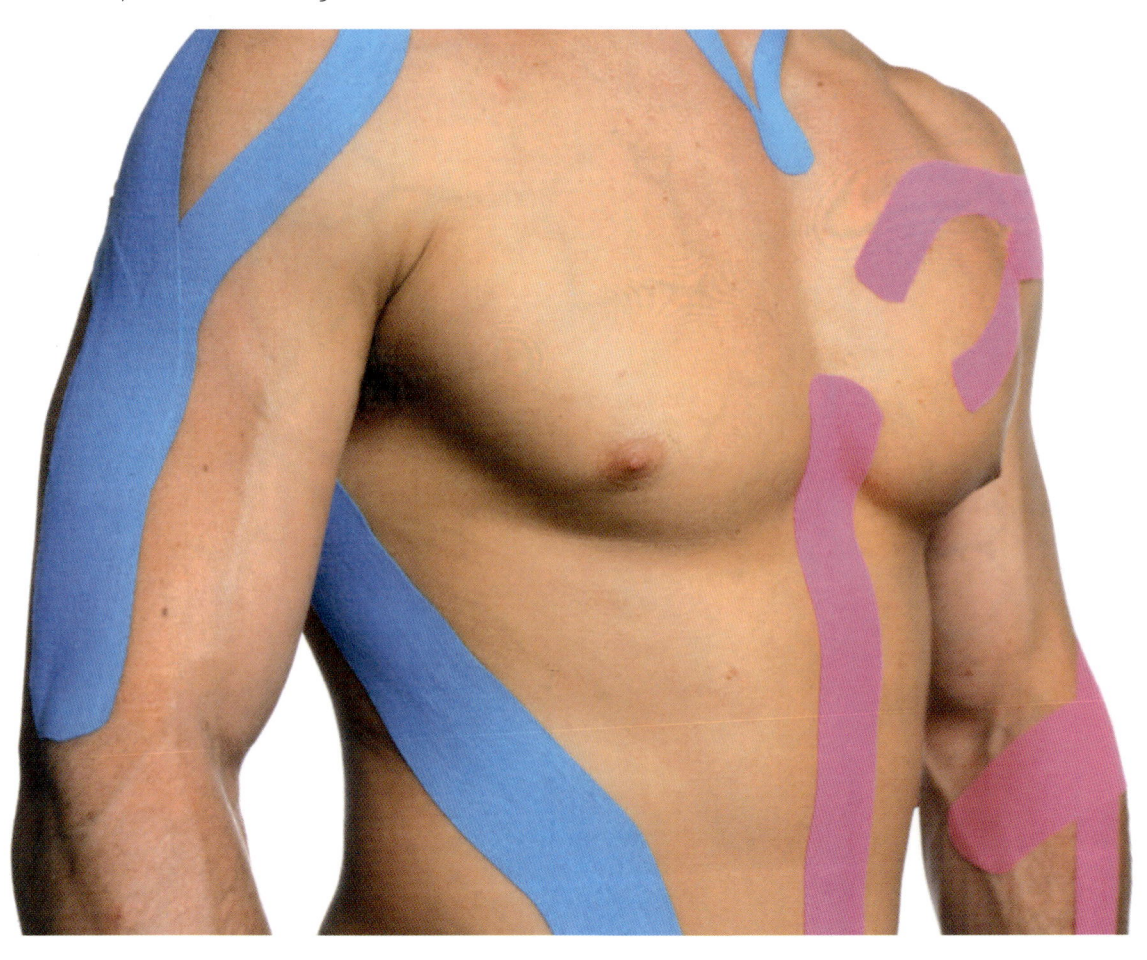

XTENSOR HAND EXERCISER

www.thextensor.com   www.amazon.co.uk   **WELLBEING**

By increasing circulation, the Xtensor targets the muscles and tendons of the hands, wrists and elbows, relieving the aches and pains that can come with working on a computer keyboard over long periods of time, whilst also building muscular dexterity for gamers. It works via resistance bands attached to each finger.

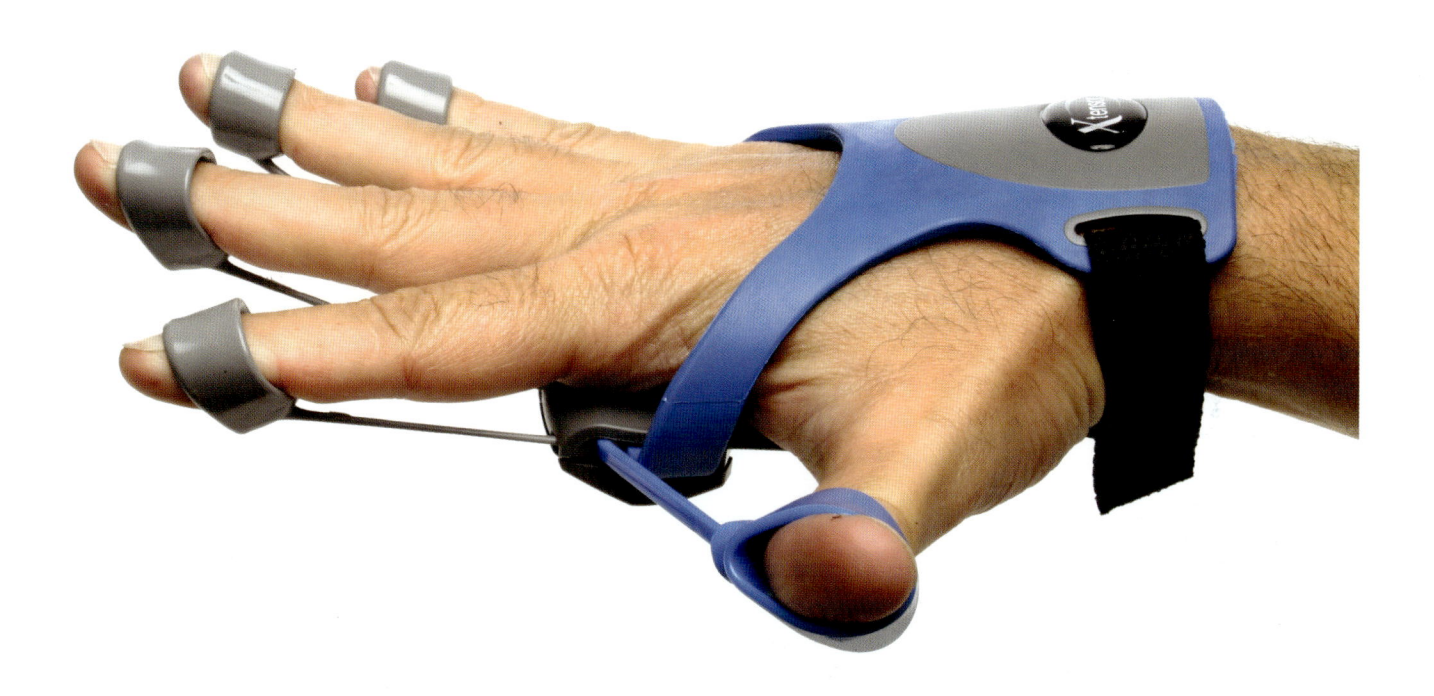

50 IDREAM 1180 HEAD MASSAGER

WELLBEING | www.dinodirect.com | www.tiao.fr

By reproducing the sensation of a hand massage and stimulating blood circulation, this futuristic-looking helmet helps to relieve migraines. It incorporates two massaging functions: vibration and electro-luminescent lamps. The beneficial effects of the helmet can be felt after a recommended period of between 5 and 15 minutes use.

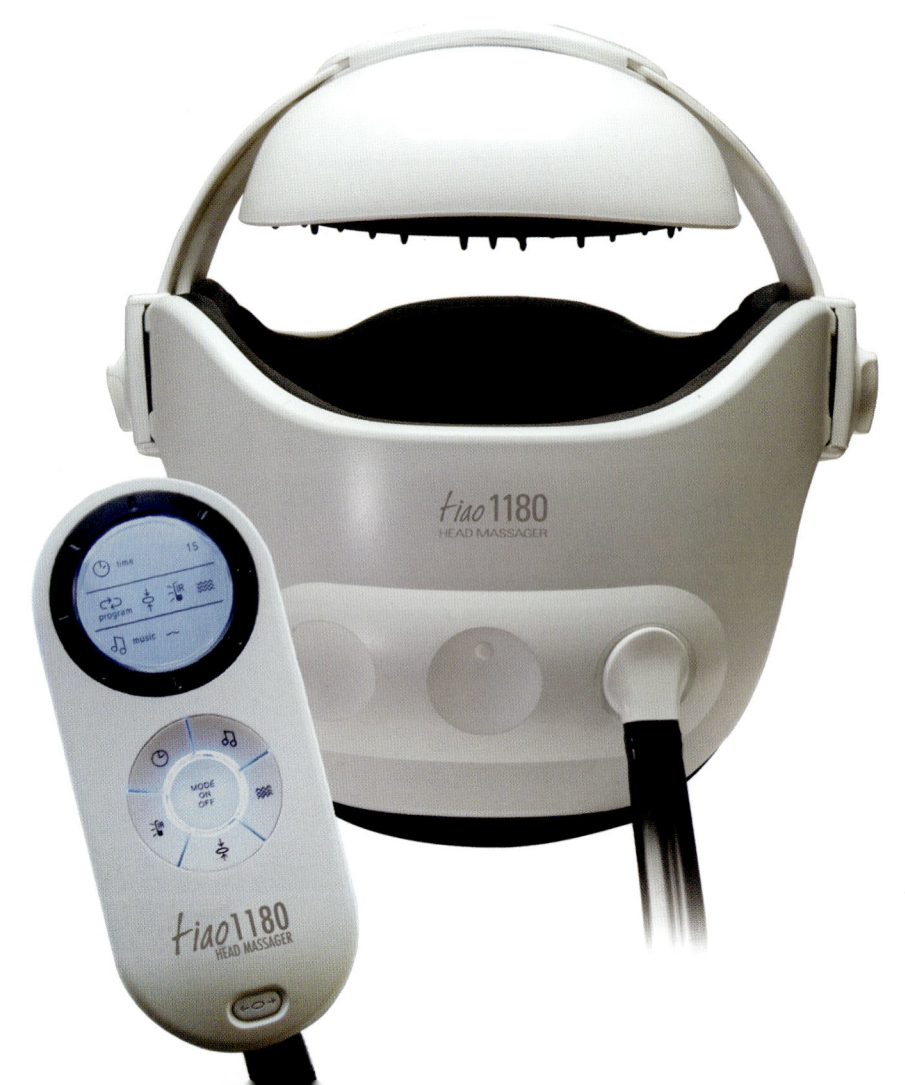

BUZZ MINI PERSONAL MASSAGER

www.brookstone.com

WELLBEING

Portable and discreet, the Buzz Mini Personal Massager takes the hard work out of a self-administered massage. There are no switches or buttons – simply apply light pressure to activate any of the three vibrating nodes, and then work into the neck, shoulders and back to loosen any stress knots. You'll be an expert before you know it.

52 UNDERWATER LIGHT SHOW

WELLBEING | www.amazon.com | www.firebox.com

Take the fun of the disco into the bathtub (leaving the spandex jumpsuit behind, naturally) with this totally watertight lamp. Ingeniously, the lamp set floats, projecting hundreds of rainbow coloured light spots onto the surface of the water. More adventurous types can even use it in the outdoor pool.

KISKIN BIKINI

www.kiskin.it

WELLBEING

So you're trying to persuade the woman in your life to get a tattoo, ideally something sexy. But she's ruining your plans by whining about needles, blood and the like. Give her a KiSkin Bikini. After a stint on the sun-lounger, the small hole, placed judiciously at the top of the bikini bottom, creates a heart-shaped suntan mark just below the small of the back.

54 — LOVE LIGHT GLOWING CONDOMS

WELLBEING | www.condomjungle.com | www.condoms.co.uk

"Surprise!" Hmm, probably best to warn the love of your life that you're trying out phosphorescent condoms before turning off the lights. They charge up with a minute's exposure to light, and then duly glow for the following ten minutes, which may or may not (now, be honest) be enough to see you through. Make up your own jokes about light sabres here.

KAMA SUTRA CONDOMS

www.atypyk-e-shop.com

WELLBEING

Even the most imaginative soul can run out of ideas in the bedroom – and it's not like you can just ask anyone for advice, never mind a diagram. So, Kama Sutra condoms could just help you out of (or even into) a tight spot. Sold in packs of five, it may take some time to work your way through all 64 of the suggested Kama Sutra positions.

56 LOVE PATCH

WELLBEING | www.atypyk-e-shop.com

Surely modern medicine should have found a cure for the broken heart by now. It's not like it's a new problem. Perhaps, though, the treatment is simpler than we might imagine. Forget wild rebound mistakes, moping about in your pyjamas and watching soppy films – maybe all we need is a simple patch. Come on, put on a patch and think happy thoughts.

KAMA SUTRA BEDSIDE BOX

www.amazon.com   www.amazon.co.uk   **WELLBEING**

A little box of tricks to keep by the bedside, ideally within reach, when you need to spice things up between the sheets. The Kama Sutra Bedside Box includes a massage cream, a few feathers to tickle the imagination (and anything else you fancy), a self-heating sensual oil and even some fruit-flavoured honey dust. Lie back, relax and indulge.

FASHION

3

60 X-RAY BAG

FASHION

www.thinkgeek.com

www.mycoolgadget.co.uk

A revolver, a loaf of bread, a tin of baked beans, a dagger? What kind of party are you going to? This cunning X-Ray Bag, with its bizarre selection of contents, seems sure to throw any would-be inquisitor off the scent. Made of flexible, woven plastic, it really comes into its own at the airport (just don't count on the security staff seeing the funny side).

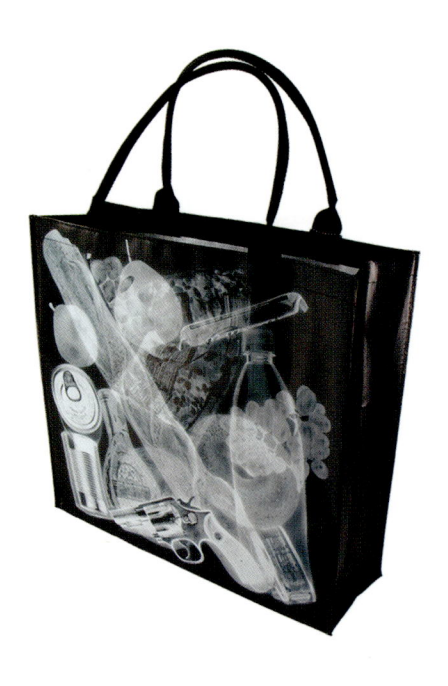

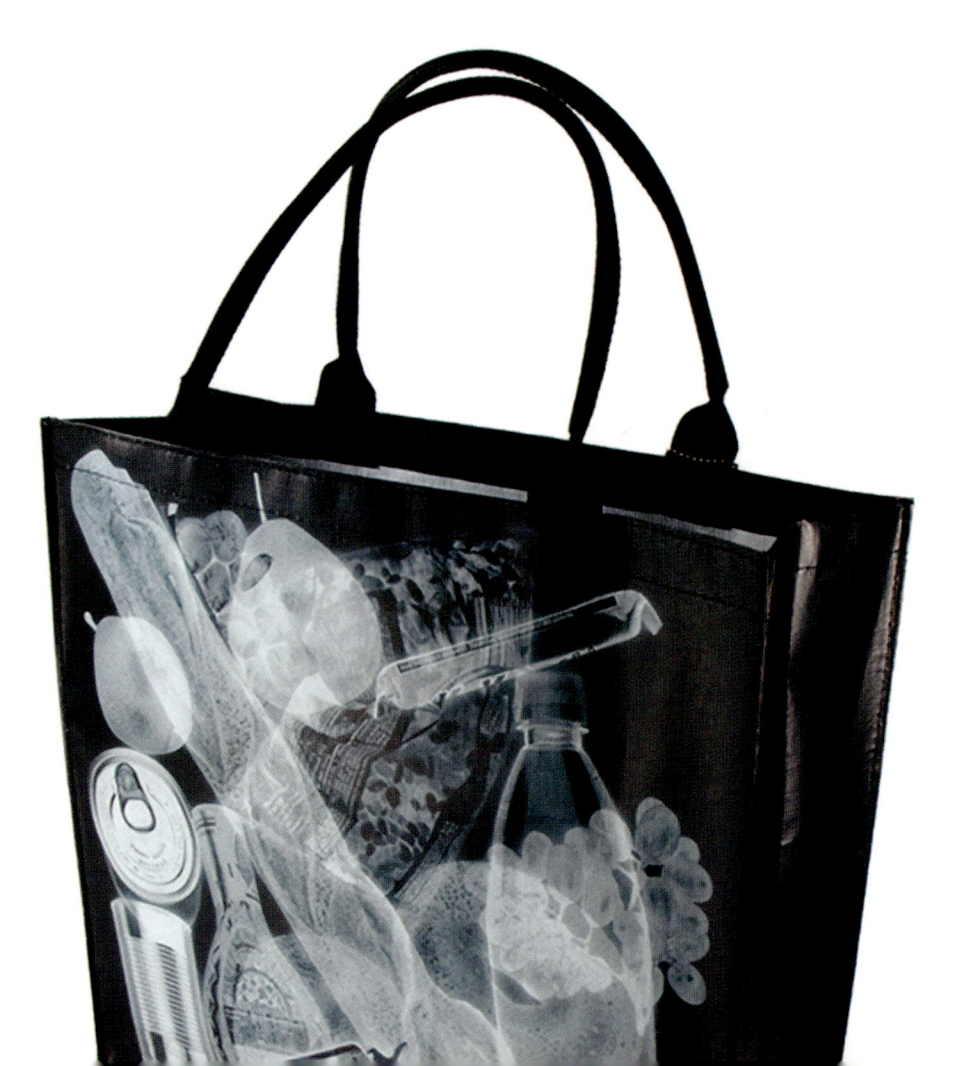

CASSETTE TAPE TOTE BAG

www.thinkgeek.com   www.giftzone.co.uk

FASHION

61

Remember the days of the tape cassette? The joys of recording from the radio or futilely spooling disobedient tape back into the case with a pencil. This Cassette Tape Tote Bag brings it all back in oversized cool. How appropriate that your new-fangled MP3 player, with its collection of 3,000 or more songs, will probably be hidden somewhere inside. Very retro and very hip.

62 LUMINOUS CLOTHES

FASHION — www.lumigram.com

Light up your next social event like never before – literally! Made with a combination of fibre optics and textiles, luminous clothes are true to their name, radiating light via the means of a small electronic module. It's all perfectly safe, and guaranteed to make you the star attraction for the evening.

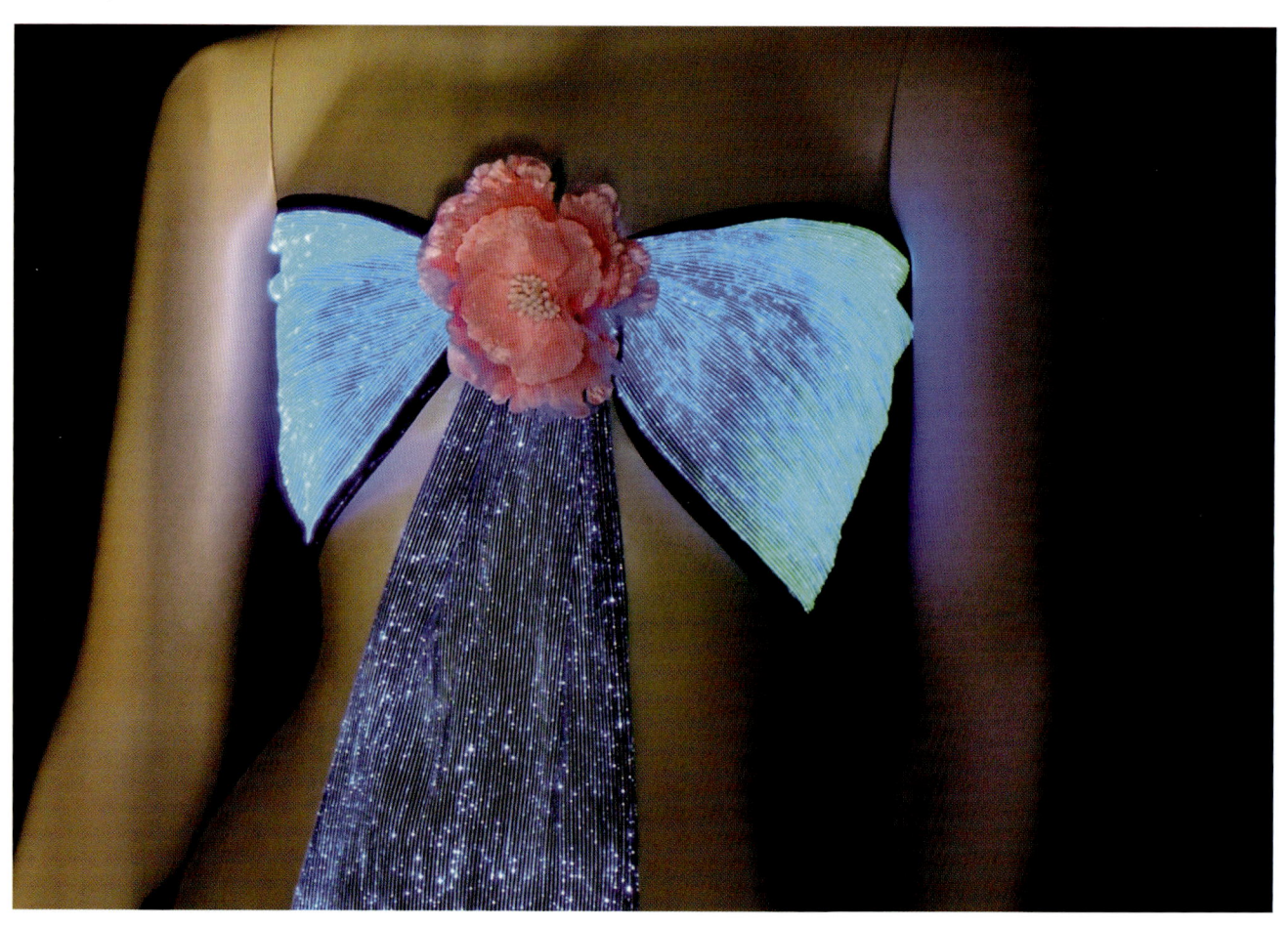

SMOON BIJOU

www.gnr8.biz   www.cosmoligne.com

FASHION

63

Simple, pure design from the Night Bijoux collection, the Smoon necklace takes on the appearance of the moon, subtly accentuating the beauty of the wearer and bringing an original and romantic touch to any evening outfit. Handcrafted from silicon, elegantly polished and linked by a silver thread to a replaceable battery, the luminous jewels are truly unique. Available in five colours.

64 COMIC BELTS

FASHION | www.bigbadtoystore.com | www.passion-estampes.com

The artistic merit of the humble comic strip has been getting the attention it deserves in recent years, marching triumphantly into the realm of cinema and repackaged for a wider audience in the guise of the graphic novel. Now, a new accolade: the comic strip belt. You can't deny it looks very cool indeed.

RECYCLED JEWELLERY

www.annabuilt.com

FASHION

Recycling meets fashion in Anna Built's jewellery. Made from reclaimed materials, often rescued from the trash and transformed into something that transcends their former identity, necklaces, earrings and other accessories carry the decoration and motifs of their old lives. Good for the planet and good for the wallet too.

THE DAMNED

FASHION — www.srulirecht.com

When is a hanky not a hanky? When it's a bullet proof guardian angel that's going to save your life. Worthy of James Bond no less, The Damned is made from lemon Military Grade Ballistic Strength Aramid Fibre, which will, once carefully folded seven times, apparently stop a bullet in full flight. We don't recommend you put the theory to the test — just take our word for it.

KOFFSKI

www.charlesandmarie.com   www.koffski.com   FASHION

All men are familiar with trying to cram a wallet, keys, change and more into their pockets. By the time the mobile phone is shoved in, there's a bulge in an awkward place. The Koffski is big enough to hold all the essentials with style. It comes with a detachable handle and can be worn discreetly tucked away or left on show; around the waist or over the shoulder.

68	**BEER HOLSTER**	
FASHION	www.gifts.com	www.holsterup.com

Ideal for the barbecue king or the dedicated angler, the Beer Holster keeps your favourite tipple exactly where you need it – on the hip. A secondary hook ensures you don't spill a drop, whatever you might be up to.

BEERBELLY

www.baronbob.com www.thebeerbelly.com

FASHION

I hope the irony isn't lost on you. It's a fake, removable beer belly used for the easy carriage of, yes, beer. The transportable thermos, capable of holding up to three litres of liquid, is worn beneath clothing to give that genuine beer gut look. A tube and mouthpiece allow you to drink direct from the belly. Of course, you don't have to fill it with beer — soft drinks work just as well.

70 BLACK SMOKING MITTENS

FASHION

www.fridgedoor.com

www.suck.uk.com

With the ban on lighting up in public places leaving smokers out in the cold, Tobi Wong has come up with an ingenious way of ensuring that you don't pay for the nicotine hit with a bout of frostbite. The smoker's mittens are specially adapted with a metal eyelet, so that you can puff away without getting cold hands. Now you just have to figure out how to get a cigarette out of the packet.

MUSUC'BAG

www.selkbagusa.com   www.musucbag.com   FASHION

The Musuc'Bag is a padded suit with built-in arms and legs to allow maximum mobility and comfort. The traditional sleeping bag has always been rather restrictive, claustrophobic even, but with Musuc'Bag you can actually get up and walk about. Perfect for frolicking around the campsite without risking hypothermia or for simply wearing around the house on a cold winter's day.

SENZ UMBRELLAS

72 | FASHION

www.senzumbrellas.com

Small, stylish, easy to use and, above all, built with a toughness that belies its modest scale, this asymmetrical umbrella can withstand winds of up to 65 kilometres an hour. The Senz Umbrella's aerodynamics help you find the best position for walking in the wind. Those days of wrestling with an inside out umbrella on the seafront, like a Mary Poppins gone wrong, might just be a thing of the past.

FUCK THE RAIN UMBRELLA

73

www.geekandstyle.com   www.artlebedev.com   **FASHION**

Rain doesn't respond to polite persuasion. Sometimes you just have to be a bit more, well, more aggressive. So, bring on the Fuck the Rain umbrella with its telescopic handle, clasp system and unashamedly rude mission statement. Created by the Art Studio, this brolly could find you praying for rain just to show it off. One piece of advice – point it skyward to avoid offending passersby.

74	**8-BIT DYNAMIC LIFE SHIRT**
FASHION	www.thinkgeek.com

Wearing your heart on your sleeve is so 20th century; these days you should be wearing your heart – all six of them – on your chest. These ingenious t-shirts, with batteries concealed in the seam, light up or fade depending on your whereabouts. Sold in pairs, the t-shirts respond to each other: as you move further apart the hearts go out; as you move closer they light up again.

PERSONAL SOUNDTRACK SHIRT

www.thinkgeek.com

FASHION

Nothing says 'I've arrived' like a t-shirt with a built-in loudspeaker playing your favourite music. Perfect for making a dramatic entrance into a packed room, or for simply keeping yourself entertained whilst cleaning the house. Complete with a remote control, built-in musical effects and the option of connecting an audio player to the battery case, finally your life has a soundtrack!

DATING SCENE CUFFLINKS

FASHION — 76

www.amazon.com www.diamondharmony.com

All those awkward first conversations: Does she? Is he? Can I? The Dating Scene Cufflinks will help get your message (and status) across without the agonising verbal ordeal of a new encounter. Each denotes a particular status: married, single, divorced, footloose and fancy free, curious, bi, hetero, desperate, etc. Indispensable for any shrinking violet.

GLOVERS

www.cosmoligne.com   www.radius-design-shop.de

FASHION

You want to be romantic, you want to hold hands, but isn't it a bit cold for all that? Here's an ingenious, original solution – Glovers are cosy and comfortable double-handed gloves. So, at last the romantics of this world can stroll hand-in-hand whilst the bitter east wind blows, safe in the knowledge that love (and the feeling in your fingers) will survive the cold.

GET NAKED BIKINI

FASHION — www.racheshop.de

"Come in, come in, the water's lovely." Try and disguise your eagerness as you usher the lady in your life to the water's edge in the knowledge that she's wearing the Get Naked Bikini, a two-piece that dissolves in water in under five seconds. And then, once in, your next – equally difficult – task is to persuade her to get out. There's also a model for men, so that she might wreak her revenge.

SOLAR SAFARI COOL HAT

www.amazon.com | www.laliquidators.net

FASHION

This cool linen hat comes with an integrated fan, powered by miniature solar panels. When the sun goes in, a back-up battery pack ensures that the air keeps on circulating. Made from waterproof linen, the hat should even keep you cool in the midst of a tropical storm. Safaris need never be sweaty again, nor for that matter should camping, fishing, golfing, hiking, bird watching...

80 — FASHION

DANCE SHOES

www.com-pa-ny.com

The days when the kindly aunt or uncle had to grin and bear the hard-to-master niece/nephew-dancing-on-my-feet ritual may finally be over. Finnish designers Song and Olin have crafted these shoes specifically designed for the job. Wear these and there's no danger of the child being lifted off their feet. So, young lady, what will it be: A waltz? The foxtrot? The Funky Gibbon?

THUDGUARD HEAD GEAR

www.babysfirstheadgear.com www.thudguard.com **FASHION**

Designed by Kelly Forsyth-Gibson, the Thudguard Head Gear gives parents piece of mind. It's a protective foam helmet that will help shield your child's head (forehead, sides and back of the head) from those inevitable bumps and bangs. An elastic strap allows you to adjust the size as the baby grows. It also meets the British Association of Accident and Emergency Medicine Safety Standards.

82 — PET COSTUMES

FASHION

www.buycostumes.com

www.costumiers.co.uk

It's fancy dress night and Fido has nothing to wear. Well, how about slipping him into a Darth Vader costume, an outfit that will enable your little pooch to explore the 'dark side of the force'? Unless, of course, he's tempted by the Princess Leia garb with its dainty headphones, or perhaps even the Yoda costume, ideal for the more mature, worldly-wise dog.

DOGGLES

www.doggles.com   www.petplanet.co.uk

FASHION

It's come to something when your dog looks cooler than you do. Little Benji is already decked out in hat, coat and shoes, so I suppose it's only natural that he should want to accessorise with a pair of sunglasses. Doggles are the first and only eye protection created specifically for dogs.

GAMES & TOYS

4

86 | GAMES

ANIMAL HANDS TEMPORARY TATTOOS

www.iloveuma.com www.npw.co.uk

And then, beneath the terrified gaze of Giraffe-hand, Crocodile-mitt pounced ferociously on Zebra-paw and gobbled him up! Designer Héctor Serrano drew inspiration from traditional children's hand games when he conceived Animal Hands Temporary Tattoos. They're removable tattoos, applied to the skin using a little water. Instantly, children have a whole jungle of stories at their fingertips.

SQUASH-PLUSH TOYS | 87

www.roadkilltoys.com

GAMES

All in the worst possible taste, the Roadkill range of Squash-Plush, distinctly uncuddly, toys give vent to the inner squirrel, hedgehog and rabbit – literally. Blood and guts spew forth from the animals, all freshly peeled from the tarmac. Your carcass of choice even comes in a zipped transparent bag complete with death certificate.

88	**HILARIOUS COSTUMES**	
GAMES	www.buycostumes.com	www.anycostume.co.uk

Who doesn't like to dress up once in a while? Pink gorillas, man-sized parrots, cowboys, croupiers – this website supplies more than 130 outfits. Some are outrageous, some are sexy; nearly all will raise a laugh. Perfect for New Year's Eve, birthdays, stag dos…

TOM ARMA BABY COSTUMES

www.halloweenexpress.com

In the full knowledge that your bouncing bundle of baby joy may well grow up into a teenage monster, get your revenge in first by dressing them up in a cute Halloween baby suit. That way, when the time comes, you'll have photographic evidence of just how cute they once were; and they, it goes without saying, will cringe with embarrassment – mission accomplished.

90 UBERARC 1600

GAMES

www.amazon.com www.uberarc.com

The UberArc 1600 puts budding architects through their paces as they construct a 1600-piece skyscraper. They have the chance to choose the location (taking care to avoid seismic zones, naturally), work on the foundations and even fill in a building permit. The kit allows them to build representations of existing skyscrapers or to design their own. Fun for all the family.

20-IN-1 MULTIPLAY GAMES TABLE

91

www.allgamestables.com   www.paramountzone.com

GAMES

Why have one games table when you can have 20? Table football, ping-pong, pool, chess, skittles, glide hockey – the 20-in-1 Multiplay Games Table has all these and more. All of the accessories and instructions required to play each game are included, and each of the 20 games packs away inside the light wooden frame when not in use.

92 STRAIGHT UP CHESS BOARDS

GAMES | www.straightupchess.com

Forget leaning forward, hunched over the chessboard in concentration, your posture suffering the effects. Instead, look straight ahead, make your move and stand back to enjoy the modern, stylish look of the Straight Up Chess Board, designed to brighten up walls rather than clutter up coffee tables. Unfettered by time limits, this game is played as and when you happen to pass by.

RUBIK'S MIRROR BLOCKS CUBE

93

www.thinkgeek.com

www.amazon.co.uk

GAMES

A Rubik's Cube in which the blocks are all the same colour — it sounds easy but is actually deceptively difficult. What they lack in colour, the blocks make up for in size variation. Move them around to create weird sculptural shapes, and then weep with frustration as you try and arrange them back into their original neat formation.

6-IN-1 SOLAR ROBOT KIT

GAMES

www.stevespanglerscience.com

www.lazyboneuk.com

A genuine toy for the green age, the 6-in-1 Solar Robot Kit gives children a first taste for the potential of solar energy. Six different models, including solar cats and dogs, a solar windmill and a solar aeroplane, are all powered by the sun's rays. There's even a model wind turbine. The models will also work under the light from a lamp.

CUBE WORLD BLOCK BASH

95

www.amazon.com   www.firebox.com

GAMES

The stick men of Cube World are getting cabin fever, stuck in their small boxes. So, connect one cube up to another cube and they can interact. They'll thank you by doing all manner of interesting stuff: playing rugby, riding rodeo and straddling monocycles. Some will even move from one cube to another. Shake the cubes or move them around to get the stick men moving.

96 INFLATABLE GLADIATOR COMBAT SET

GAMES | www.find-me-a-gift.co.uk | www.gadgetepoint.co.uk

"I simply said, 'Come on love, let's talk this over like reasonable, civilised adults'. And then she just flew at me with an inflatable mace." Why talk things out when you can attack each other with this blow up Gladiator Combat Set, complete with shield, helmet and aforementioned spiked club? Perfect for letting off steam or settling old scores without the downside of physical pain.

FUNTRAK PAINTBALL PANZER

www.funtrak.co.uk

GAMES

Always wanted your own tank but been held back by cost, size, implications for world peace, etc.? The Funtrak Paintball Panzer could be your way in. It's a genuine tracked vehicle, with a fully integrated paintball gun that allows for 'firing whilst driving'. Built to order, the paintball panzer is available in a range of colours from khaki to sky-blue.

98 ROLERBALL

GAMES

www.iwantoneofthose.com   www.rolerball.co.uk

The Rolerball doesn't actually make a noise, but if it did it would probably be something like "Aaarrgghhh". Harnessing the participant inside the giant transparent sphere with two straps, it makes the most of gravity and sends you crashing (safely) down to the bottom of the hill. One for the extreme sports enthusiast. Hill not included.

VAMPIRE 3 WINGSUIT

www.phoenix-fly.com   www.opulentitems.com

GAMES

A wingsuit that prolongs the glide time (and adrenaline rush) for freefall parachutists and base jumpers. The tailor-made suit gives precise control and high forward speed, generating longer flights and greater freefall time. Not for the faint hearted, nor the novice — only experienced skydivers need apply.

100 MONKEY LIGHT BIKE LIGHT

GAMES | www.amazon.com | www.monkeylectric.com

Don't just 'be seen' on your bike at night, go the extra mile and put on a fantastical light show. The Monkey Light uses 32 LEDs to transform your pedal power into spinning wheels of light. Easily mounted on the spokes of nearly any bike wheel, the lights can be programmed to mimic a neon advert, display an illuminated message or recreate the dazzle of a fairground.

THE MAGIC WHEEL

www.amazon.com
www.magicwheel.co.uk

101

GAMES

Somewhere between the skateboard, the scooter and the monocycle resides the beguiling Magic Wheel. Put your foot on one side of the wheel, push off with the other, and away you go. To brake, simply put your foot back down on the ground. Sleek, cool, environmentally friendly and easily mastered.

102 | GAMES | THE WORLD PUZZLEBALL (540 PIECES)

www.amazon.com · www.amazon.co.uk

If you thought reconstituting the Horn of Africa was difficult, wait until you start piecing together the Pacific Ocean. The World Puzzleball, comprising 540 pieces, is what you might imagine it to be — a puzzle in globe form. No pressure, but the original took seven days to create — that should give you something to aim for.

LEVITRON ANTI GRAVITY EARTH GLOBE

103

www.innovatoys.com | www.amazon.co.uk

GAMES

Like a real Earth in miniature, the Levitron Anti Gravity Earth Globe turns and hovers in space. A simple system of magnets achieves the desired effect. The four-inch globe and beautiful chrome base will look incredible on any desktop.

104 FLYING F*CK HELICOPTER

GAMES www.thinkgeek.com   www.thumbsupuk.com

Remember the olden times, when you dispatched messages via a homing pigeon, attaching a piece of parchment to spread the latest news? Well things have moved on a bit since then. Moved on so far in fact that you can now send a colleague or friend an airborne 'F*CK'. This helicopter does actually fly; just make sure you direct it at the right person.

FLYTECH DRAGONFLY

105

www.amazon.com | www.amazon.co.uk | GAMES

The first radio controlled insect no less, the Flytech Dragonfly possesses excellent stability in the air, lands softly, climbs, dives and hovers. The Dragonfly is built with a durable carbon-fibre structure, and is duly light enough not to damage the inside of your house. It even flaps its wings like the real thing.

106 FLYBAR POGO STICK

GAMES   www.flybar.com   www.amazon.co.uk

Behold, the next generation of pogo sticks. The Flybar uses an ingenious piston system that can propel the user to heights of 1.5 metres. As you become more proficient, the versatility of the Flybar means you can learn a few tricks. Equally, the adaptability of the piston system allows you to adjust the desired level of bounce.

FREELINE SKATES

www.freelineskates.com   www.driftingskates.com

GAMES

Freeline skates comprise one deck for each foot. They're ridden skateboard/snowboard style but the feet aren't fixed, with movement achieved by rotating the skates. You know you're in business when you start making smooth 'S' shape turns riding downhill. Sounds tricky, but they're more intuitive than roller skates and provide a great introduction to the wider world of board sports.

108 CONTROL-A-CAT REMOTE CONTROL

GAMES www.thinkgeek.com www.npw.co.uk

Dogs do needy; cats do whatever they want – 'twas ever thus in the animal kingdom. But what if a remote control could get Kitty under control? What if you could just point, press the appropriate button and suddenly the feline spirit was tamed. Control-a-Cat Remote Control could do the job. Focus your mind on the remote control really, really hard. Now push the button. Did it work?

CAT PLAYHOUSE

www.suck.uk.com

109

GAMES

Cats, by their very nature, seem to look bored at least 97% of the time. But perhaps they're just not getting enough intellectual stimulation. Surely a kitty-scale fire engine would sort that out? Or an aeroplane, a car or a tank. The suck.uk.com website has all of the above, delivered to your favourite moggy in flat pack form (they might need a helping paw with construction).

110

GAMES

DOGLLYWOOD DVD

www.dogllywood.com   www.gift-shop-for-dog.com

The canines of Dogllywood are more than just box office stars with lavish lifestyles. They also play an important social role: they keep your dog amused during those difficult periods when you have to leave your pooch on its own. The Dogllywood DVD includes bonus footage designed to encourage socialisation and to accustom dogs to everyday noises such as cars, thunderstorms, etc.

After the immense success in Japan and the U.S.A., now at last available in Europe!

R2 FISH SCHOOL KIT

111

www.r2fishschool.com   www.petplanet.co.uk

GAMES

Does Nemo's life lack excitement? Is there seaweed in his goldfish bowl? Is there anything to fight the boredom? Give him something to do – give him the R2 Fish School, a complete fish training kit. Before you know it he'll be scoring goals, shooting hoops, slaloming and limbo dancing. It's all done by training the fish to follow food.

GADGETS

5

114 MESSAGE TAPE

GADGETS www.thinkgeek.com   www.suck.uk.com

Cut off a strip, fix it where you want it (it's adhesive), and then black out the unwanted segments with a marker pen to make a note or write a message. Message tape is ideal for taping up boxes and labelling the contents. It can also be used to leave messages for loved ones, for teaching children to write or even as part of an art installation.

TALK TO THE HAND STICKY NOTES

www.coolstuffexpress.com   www.neatoshop.com

GADGETS

No matter how sociable – amiable even – you might generally be, some days you just want to be left alone. And nothing says 'leave me alone' better than the universal 'talk to the hand' gesture. The hand commands respect; the hand is no-nonsense. These Talk to the Hand Sticky Notes come in a handy block – just peel them off and write your message.

116 CALENDAR TAPE

GADGETS | www.clickshop.com | www.suck.uk.com

Tying a knot in a hanky lost its practicality with the advent of paper tissues. And scribbling a note on your hand is fine until sweaty palm moments arrive. Instead, if you must make notes on the hoof, write it on some Calendar Tape and stick it to the nearest smooth surface. The yellow adhesive tape comes in two parts: the dates are on one side and the days are on the other.

ORIGAMI STICKY NOTES

www.baronbob.com   www.suck.uk.com

GADGETS

117

Doodling on a notepad is all well and good, but isn't it time you negotiated those boring meetings or phone calls with something a little more creative? Recycle your post-it notes and let the wonder of origami fill the dull moments. These Origami Sticky Notes, with 100 sheets per pack, offer up a range of challenges, including butterflies, chickens and parrots. Minimum age: 8 years.

118 PEN FISHING ROD SET WITH CASTING REEL

GADGETS

www.nauticalia.com

www.amazon.co.uk

How do you catch the world's smallest fish? With the world's smallest fishing rod – obviously. This Pen Fishing Rod set has a reversible handle on the casting reel (and so can be used right or left handed) and a line brake/clutch. It's so discreet the colleagues in the office won't suspect a thing when you disappear for an afternoon-long 'meeting'.

GRASSY LAWN CHARGING STATION

www.thinkgeek.com · www.lazyboneuk.com

GADGETS

If it isn't the mobile phone that's cluttering up the workspace, then it's the camera battery charger or the MP3 player. Bring some order to the chaos, rid the desk of that entanglement of cables – use this green and pleasant charging case to reinvigorate all the indispensable little gadgets of everyday life.

TALKING PHOTO ALBUM

GADGETS | www.amazon.com | www.talkingproducts.co.uk

The digital photo frame has become a familiar fixture on the mantelpiece in recent years, but what of the talking photo album? At first sight it looks like a traditional album holding 24 photographs. However, the Talking Photo Album has no need for written captions; it allows you to record a ten second message to accompany each photo.

TRUTH WRISTBAND KIT

www.makershed.com

GADGETS

The cookie jar is empty. She's got chocolate crumbs peppered around her mouth. And yet, still, she denies all knowledge. You've got no option but to turn to the Truth Wristband Kit. Consisting of a small box connected to a sensor placed on the finger, the gadget measures the 'galvanic skin response' or 'electrodermal response'; in short, if she gets the guilt sweats the red light comes on.

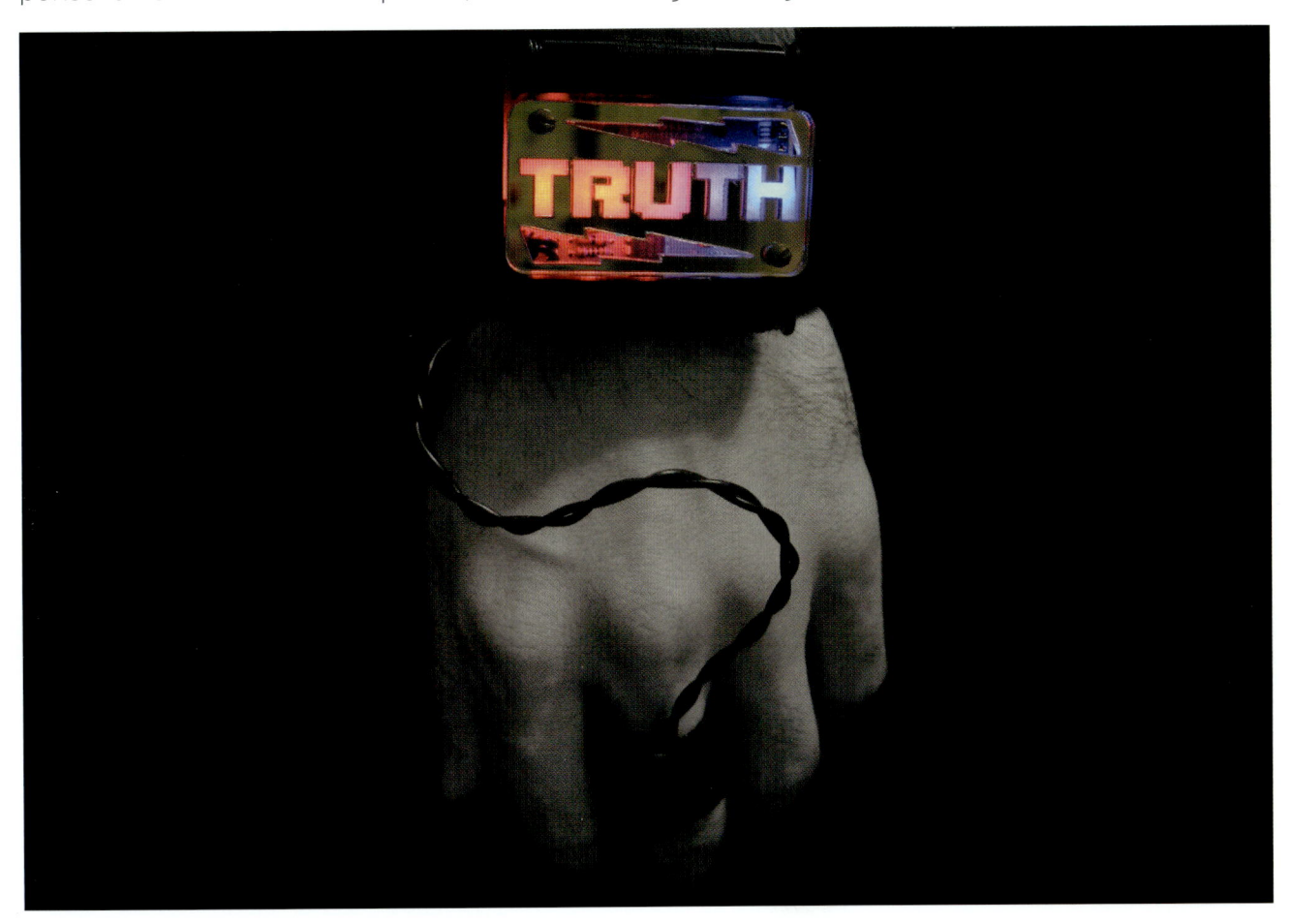

122 MIMOBOT

GADGETS

www.mimoco.com

www.pixmania.com

No, technology isn't sad, and no, geeks aren't sinister, acne-suffering loners (not all of them anyway). Computing can also be amusing, fun and even sexy, as these Mimobots prove with aplomb. They're USB flash drives cheekily disguised as our favourite heroes, from Darth Vader to Hello Kitty. Take your pick from dozens of characters.

USB HUMPING DOG

www.amazon.com

www.iwantoneofthose.com

GADGETS

First came the pyramids of Giza. Then, centuries later, the internal combustion engine and penicillin. Now, mankind moves forward once again. Behold, the USB Humping Dog. Initially, it looks like any other cute dog, but plug into a USB port and it just can't stop humping. Do we need to point out this gadget can't actually be used for storing data?

124 MUSHROOM DESK MINI CLEANER

GADGETS | www.gadget.brando.com | www.amazon.co.uk

Are you a cleanliness freak? Do crumbs make you shake with rage, or will the tiniest pencil shaving reduce you to a blubbering wreck? You need a pocket vacuum cleaner. They don't come much cuter than the Mushroom Desk Mini Cleaner. Perfect for getting dust out of a keyboard or fluff off clothing.

NYOKKI PETS

125

www.amazon.com www.find-me-a-gift.co.uk

GADGETS

Who didn't love growing cress in cotton wool when they were little? Revisit those childhood pleasures with these ceramic eggshells, each one planted with seeds that will produce a good head of grass within barely two weeks. Then you can choose whether to go for the hirsute look or to administer a haircut.

126 LOVE ME STAMP

GADGETS | www.atypyk-e-shop.com

Love letters are a great idea, it's just the washing needs doing, that child won't feed itself and you're supposed to be at work by nine. Really, who's got the time to put pen to paper? A Love Me rubber stamp simplifies the message but still retains the essence of that all-important loving sentiment.

ROSE BATH BUDS

www.iwantoneofthose.com

GADGETS

The luxurious delight of relaxing in a hot rose-scented bath isn't the sole preserve of the wealthy anymore. Anyone can do it with Rose Bath Buds soaps. Scatter the scented soap petals over the surface of the water and they melt into the bath in a few moments, releasing a delightful rose fragrance as they dissolve.

128 STITCHING POST CARD

GADGETS

www.uncommongoods.com

www.details-produkte.de

Some have said that the journey is as important as the destination, but you can't deny that travel can be a little boring at times. Relief from the monotony comes in the shape of a needle and thread with which to embroider out your travel route. Keep your mind and your fingers busy by threading your way round the continents on these pretty paper postcards.

WOODEN POSTCARD

www.amazon.com   www.suck.uk.com

GADGETS

129

Carving a message of love in wood is perhaps the original romantic gesture. But where's a tree when you need one? These thin softwood postcards provide a simple solution, offering the opportunity to write your sweet nothings the old-fashioned way. The cards are easily inscribed with a key or something similar.

130 EX-LOVER VOODOO DOLL

GADGETS   www.curiosite.com   www.gadget-box.com

Who hasn't been jilted by a lover at some point in their lives? You thought he felt the same, thought he was madly in love too. And then along comes Patricia in accounts and, well... The Ex-Lover Voodoo Doll won't bring him back, but it will absorb some of your anger. Nothing complicated: just go for it with the needles. A photo of his face stuck to the head is good for added realism.

SKY LANTERN

www.amazon.com   www.firebox.com   **GADGETS**

131

There's something truly magical about releasing a lantern up into the night sky and watching it float gently into the distance for 20 minutes or more. Made from flame-retardant paper and waxed cotton, the lanterns are easy to light and release. Perfect for a summer's evening: for weddings, parties or a romantic night on the beach.

132 PANAMAP

GADGETS — www.panamap.com

Picture the hapless tourist in the big city, rifling through multiple maps. A map for this, a map for that: confused, lost, flapping about like a penguin in the wind. The Panamap offers a compact, practical and smart solution. It's three maps in one, all printed on the same side. Information on areas, streets and the underground is printed in three layers; visible as you vary the map's angle.

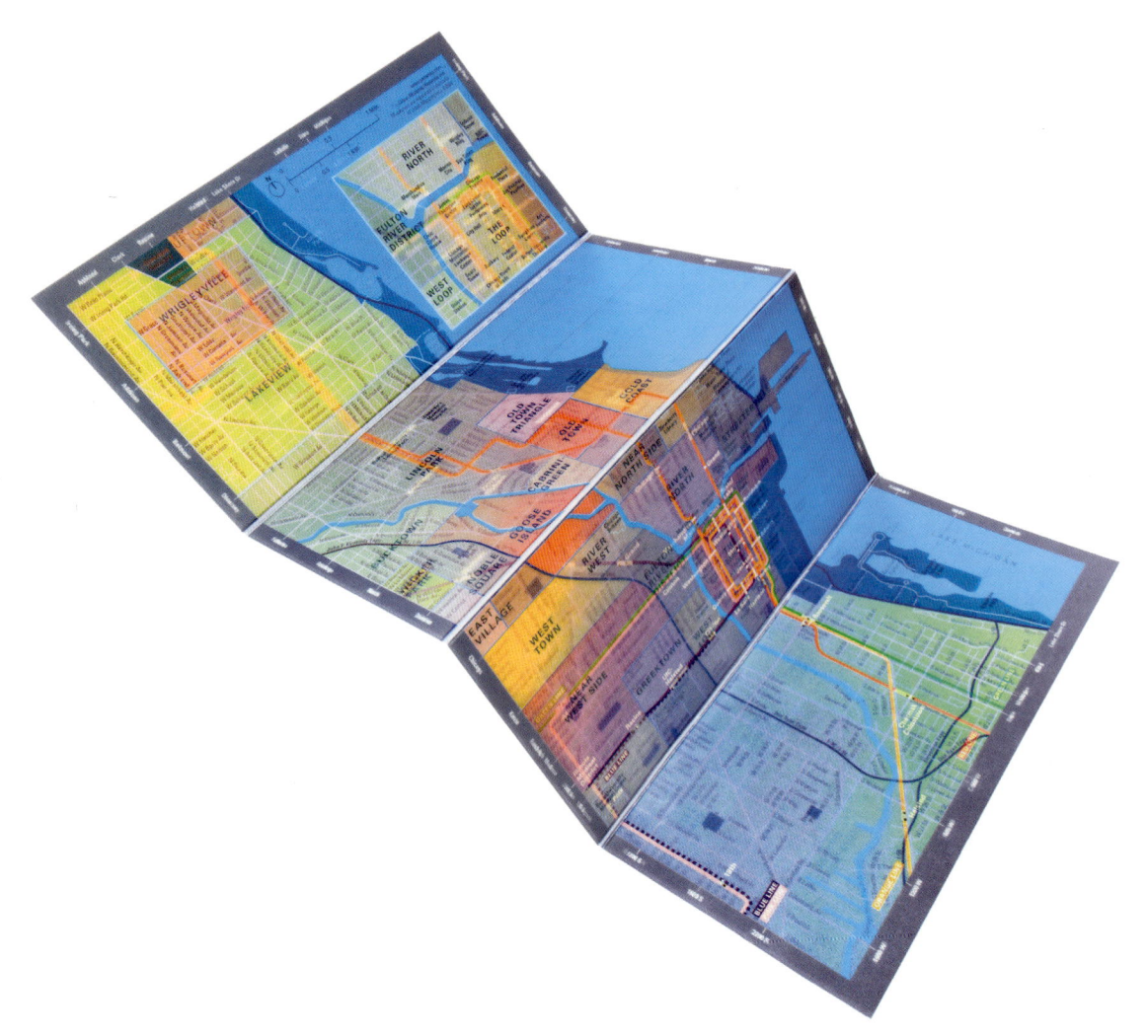

MYDOTDROPS CUSTOM TRAVEL SUITCASE

133

www.mydotdrops.com

GADGETS

A customised travel suitcase not only brings a sense of style to those jaunts abroad, it's also a boon at baggage reclaim where every piece of luggage on the carousel looks identical. It works like this: go to the MyDotDrops website, select the colours you want to use and then arrange the dots on your virtual suitcase. Your finished suitcase will arrive on the doorstep a few weeks later.

TECHNO

6

136 YOOSTAR ENTERTAINMENT SYSTEM

TECHNO — www.amazon.com — www.yoostar.com

"You got heart, but you fight like a goddamn ape." Well that's a bit harsh. Oh, I see, it's a line from the film. Yoostar puts you right in the thick of a movie (in this case 'Rocky'), without having to endure the rigours of the casting couch. Combining webcam, green screen and software, you can now place your own brilliant performance into real film clips. Watch out Sly Stallone!

MINORU 3D WEBCAM

137

www.amazon.com   www.firebox.com   TECHNO

The Minoru 3D Webcam films and displays two images, one tinted red and the other tinted blue. The two image streams are displayed on the screen but separated by a slight time lag. Slip on the famous coloured specs that come with the webcam and enjoy the perfect 3D vision. Just try and restrain yourself from reaching out to touch your webchat buddies.

138 — ECHO BOT VOICE MESSENGER

TECHNO   www.amazon.com   www.firebox.com

Practical and cute, the boggle-eyed Echo Bot comes with an important message; a message recorded by you. It's a stealth voice messenger. Simply record your message (of up to ten seconds in length) and place the robot somewhere in the house. As soon as someone passes within a metre of the eye, it pipes up with your message: "Back away from the fridge".

LIQUID IMAGE VIDEO SWIM MASK

www.divecamcentral.com   www.firebox.com

TECHNO

"Keep, glug glug, still, glug. Now, say 'cheese!'" The Liquid Image Video Swim Mask is the first underwater mask fitted with a digital video camera. Useable to a depth of 15 feet, the mask is ideal for fun in the swimming pool or for filming or photographing (it also takes digital snapshots) whilst out snorkelling or diving.

140 DIGITAL VIDEO MEMO

TECHNO www.amazon.com   www.iwantoneofthose.com

A funky 21st century take on the post-it note, the Digital Video Memo is fitted with a camera and microphone. The device can record messages of up to 30 seconds in length and offers a handy reminder of your audiovisual note by displaying a red light. Charged via a USB port, the device is magnetised and will live happily on your fridge door.

VUZIX IWEAR

141

www.vuzix.com | www.firebox.com | TECHNO

This is the stuff of science fiction: a pair of wraparound shades that plug into almost any video device and play games or films right in front of your eyes. The two LCD screens promise the viewing equivalent of a 46-inch screen as seen from a distance of ten feet. Ideal for the video gamer, the soap addict or the film buff.

142 — NOISELESS USB KARAOKE MIC FOR WII

TECHNO — www.japantrendshop.com

Karaoke's great isn't it? Yes, unless you live next to someone with their own karaoke machine and delusions of talent. But relief is at hand, courtesy of the Noiseless USB Karaoke Mic for Wii, which does pretty much what the name describes. The cone surrounding the mic quashes the decibels, yet the budget-Mariah Carey next door still gets to listen in through the headphones.

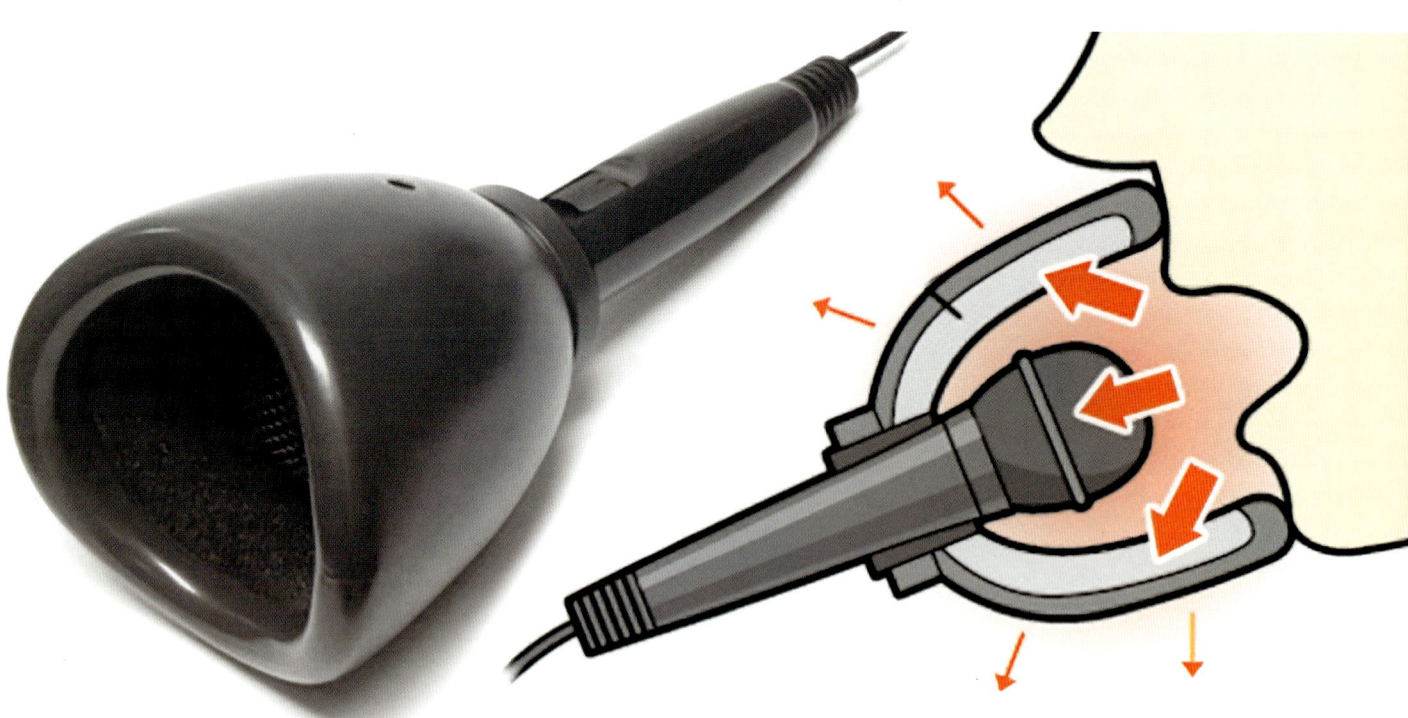

ROLLY™ SOUND ENTERTAINMENT PLAYER

143

www.sonystyle.com   www.ebay.com   TECHNO

It's the sound system with a personality: the Rolly™ Sound Entertainment Player from Sony is a robotic MP3 device that dances whilst it plays music. Twirl the Rolly to switch tunes or to hike up the volume. It even changes colour as the music varies. Bluetooth capability allows you to play music from other devices wirelessly through the Rolly.

TECHNO

144 YOROZU AUDIO SOUND REVOLUTION KIT

www.japantrendshop.com www.amazon.com

The Yorozu Audio kit lets you use virtually anything as a speaker. Take a milk carton, for example. Simply place the vibrating extension onto the surface of the carton (using the adhesive sheets included in the kit), plug in the audio, and wow, music through your milk carton! It works through the transmission of acoustic vibrations. Try it on coffee tables, lamps, suitcases, white boards, etc.

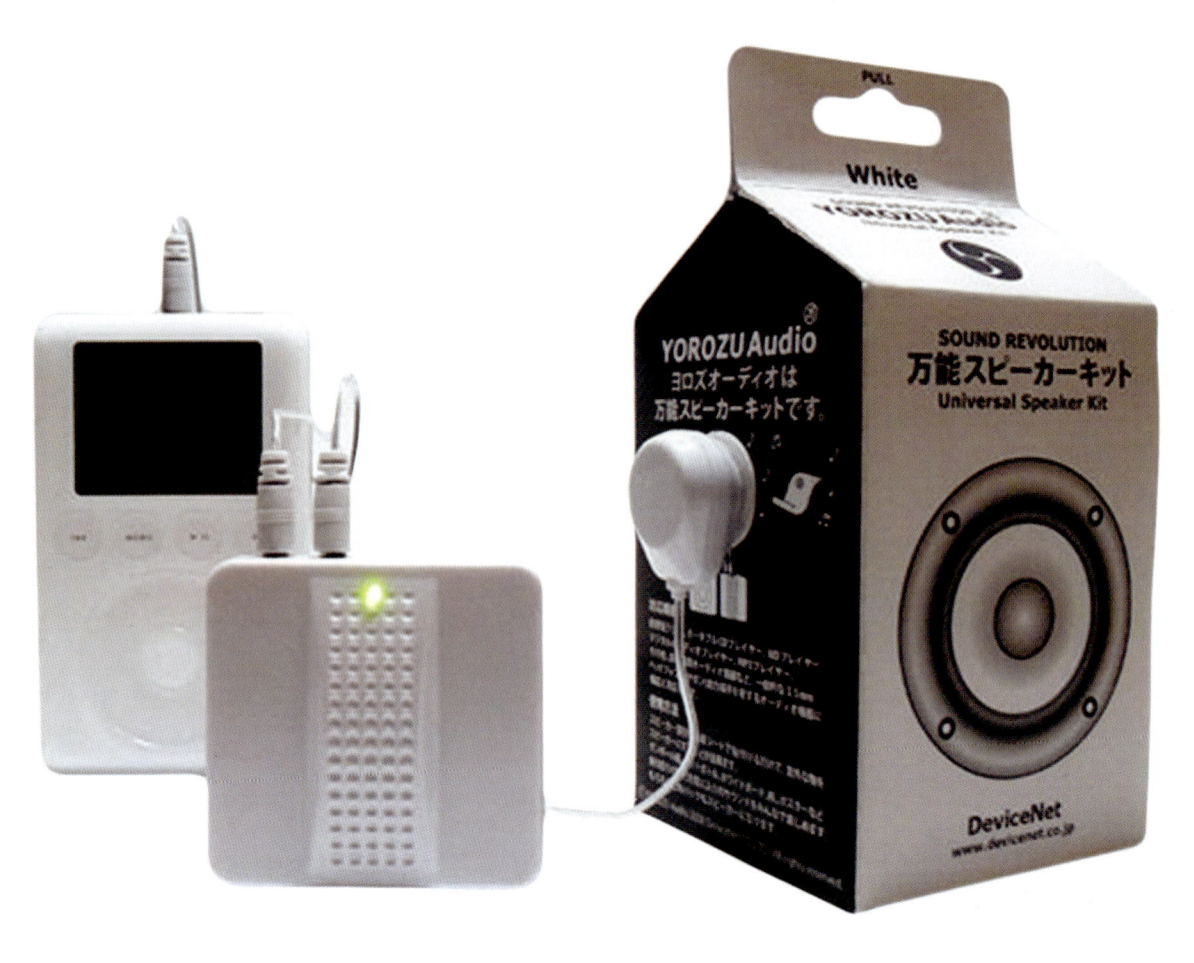

500 XL MP3 PLAYER SPEAKERS

145

www.perpetualkid.com

www.shinyshack.com

TECHNO

Contrary to what some might say, size does matter. And sometimes it's a case of the bigger the better. These giant white buds might look rather like Shrek's iPod earphones, but they're actually speakers for computers or MP3 players. The speakers have an amp built-in, and run on batteries or through your PC via a USB cord.

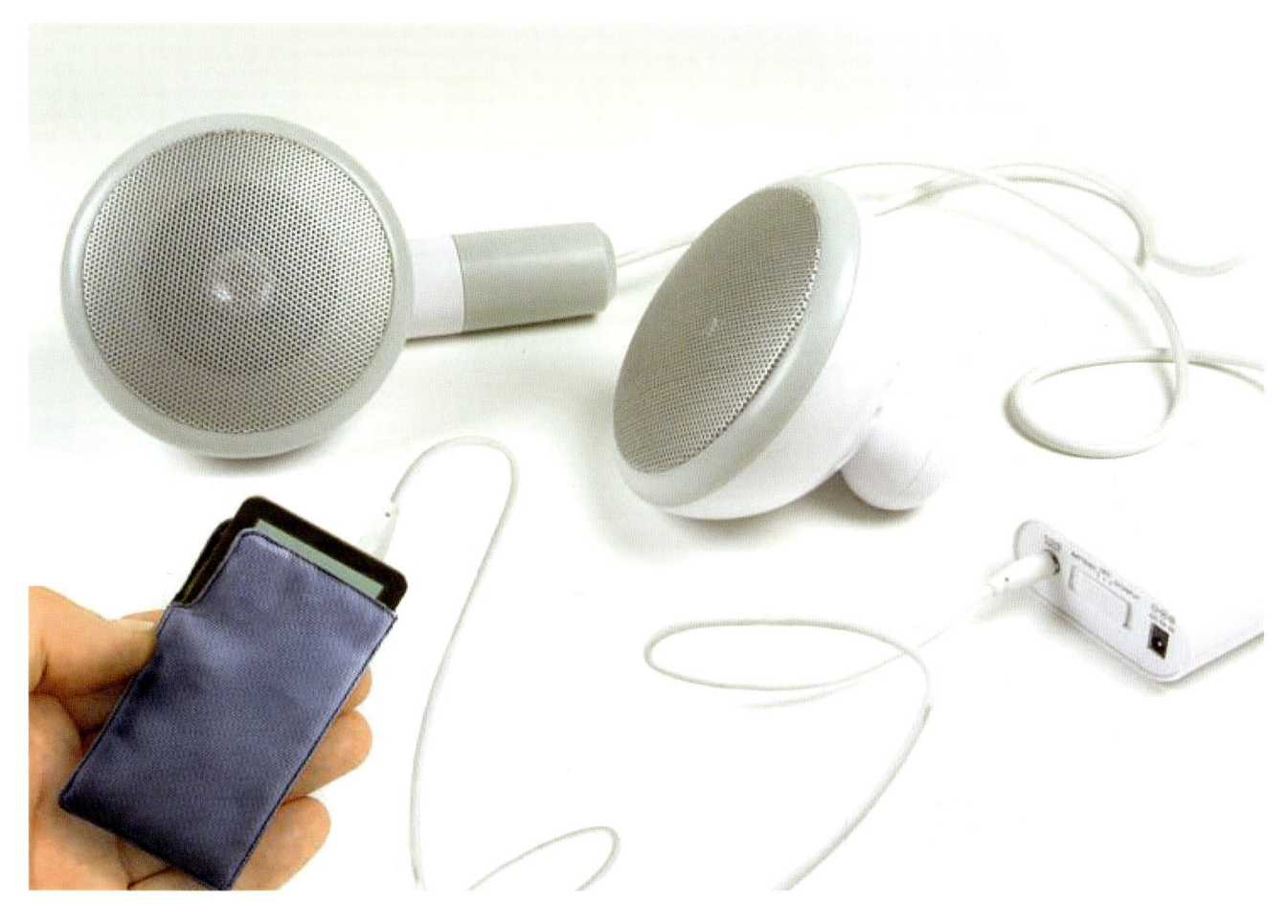

146 EMPEROR WORKSTATION

TECHNO — www.novelquest.com

With an air conditioner, air purifier, daylight simulator and touch screens, the Emperor Workstation takes multitasking to a whole new level. With so much to play with, being at work might actually be fun. It's an argument worth trying on your boss when you suggest she pays for your new workstation.

THE BEAMZ

www.thebeamz.com

147

TECHNO

Plinkity plonk, plinkity plonk. The piano's a great instrument but it takes years to learn and, frankly, who's got the time? Perhaps you should try the Beamz, an instrument played by interrupting laser beams. Each time you break the beam, a note sounds. Different song patterns (comprising 30 songs and 19 musical genres) can be followed by connecting the Beamz to a screen.

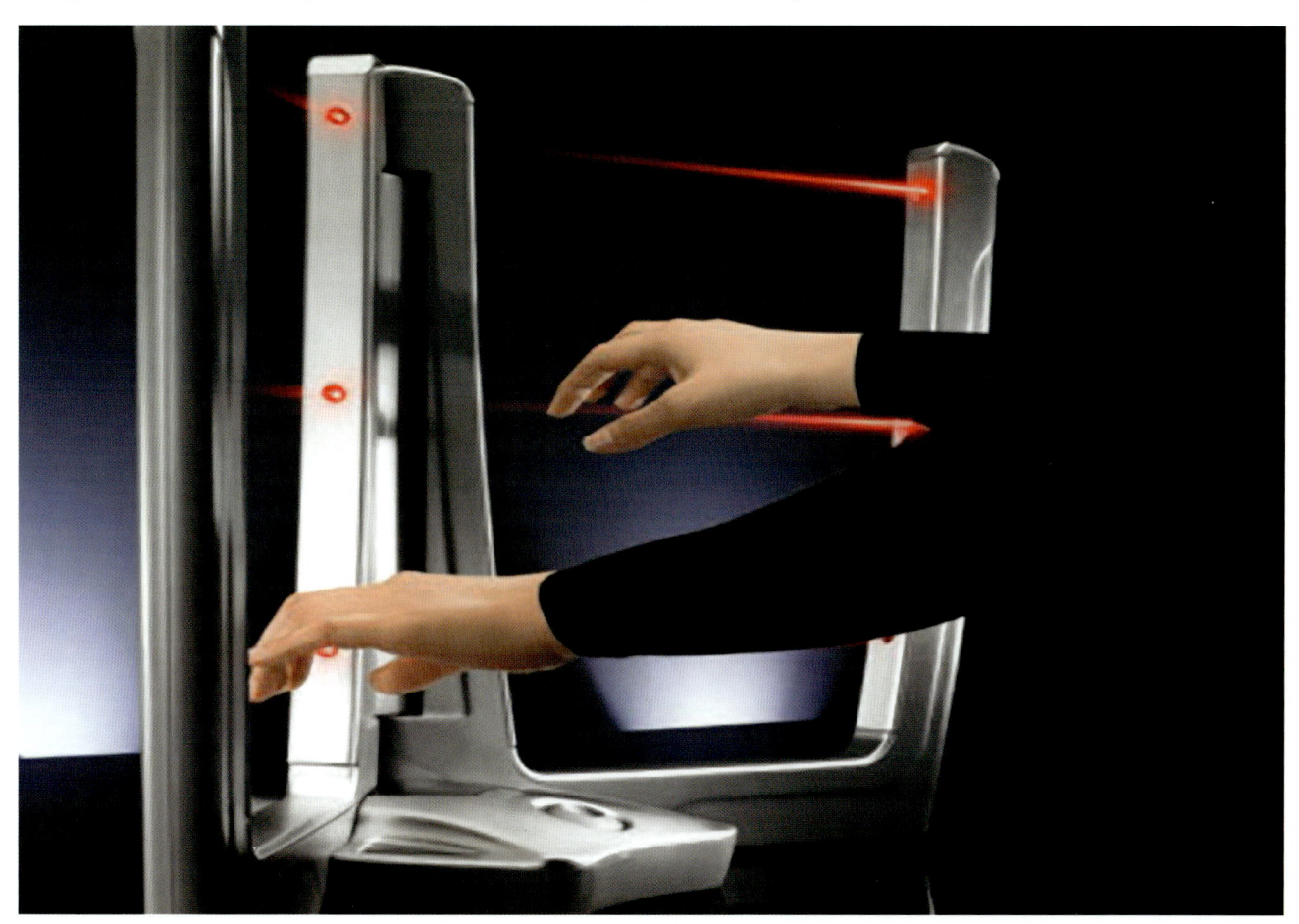

148 KEYBOARD FOR BLONDES

TECHNO www.keyboardforblondes.com

We've all got a favourite blonde in our lives — metaphorically if not actually — and it can break your heart to watch them struggling with the demands of a conventional keyboard. So, Keyboard for Blondes makes it easy: the number keys look like domino faces; the 'Return' key says 'Oops!' and 'Esc' reads 'No!'. All they've got to do now is work out how to plug it in.

LOGITECH G19 KEYBOARD FOR GAMING

www.logitech.com

149

TECHNO

The G19 Keyboard, specially devised to maximise the ergonomics for gaming, promises extreme performance. Its 12 programmable keys (called 'macros') and Colour GamePanel™ with customisable background are complemented by an extension, headphones and a laser mouse.

150 HOMESTAR HOME PLANETARIUM

TECHNO | www.japangadgetshop.com | www.amazon.com

Gazing up at Orion, Mercury, the Great Bear et al., it's hard to imagine any better feeling than being out under the stars on a clear evening. But what about in deepest winter? Mmm, perhaps not. The Homestar Home Planetarium brings the wonders of the universe into your sitting room. A special version even allows you to enjoy the constellations from your bath.

MOON IN MY ROOM

151

www.walmart.com   www.iwantoneofthose.com   TECHNO

This novel wall-mounted lamp reproduces the 12 phases of the Moon's cycle. As the room darkens a sensor is activated, which then switches off after 30 minutes. A remote control also allows you to direct the lamp manually. Perfect for the budding astronaut or for the child who's afraid of going to sleep with the light off.

152 USB TURNTABLE

TECHNO | www.usbturntables.net | www.firebox.com

Poor old vinyl. Unloved and on the shelf since the advent of the MP3 file. At least have the decency to transfer your old records to digital files before you drop them off at the thrift store. This USB turntable gives you the necessary technology to make the change. An iPod version means it can also be used as a docking station.

DISCOVER DJ

153

www.ionaudio.com   www.firebox.com

TECHNO

DJ's like David Guetta make mixing and scratching look like the easiest (and best) job in the world, but those who've done their time on the decks know how hard it actually is. Anyone with dreams of becoming a DJ should start their apprenticeship with Discover DJ. The software and turntables provide everything you need to DJ parties, events and even clubs.

154 BRIGHTFEET LIGHTED SLIPPERS

TECHNO | www.brightfeetslippers.co.uk | www.comforthouse.com

These slippers help you to potter about in the dark without fear of knocking into the coffee table or tripping over the telephone wire. Available in blue, beige or black, and in various different sizes, they're equipped with little lights that show exactly where you're putting your feet.

WATTSON

www.diykyoto.com www.firebox.com

155 TECHNO

At a time when it's cool to be green, the Wattson unit keeps track of your carbon footprint. The device monitors how much electricity is being used in your home, school or office. The greener you're being, the bluer the Wattson turns; and then, as your electricity consumption rises, it goes red with anger (giving detailed energy readings as it does so). The unit is cordless.

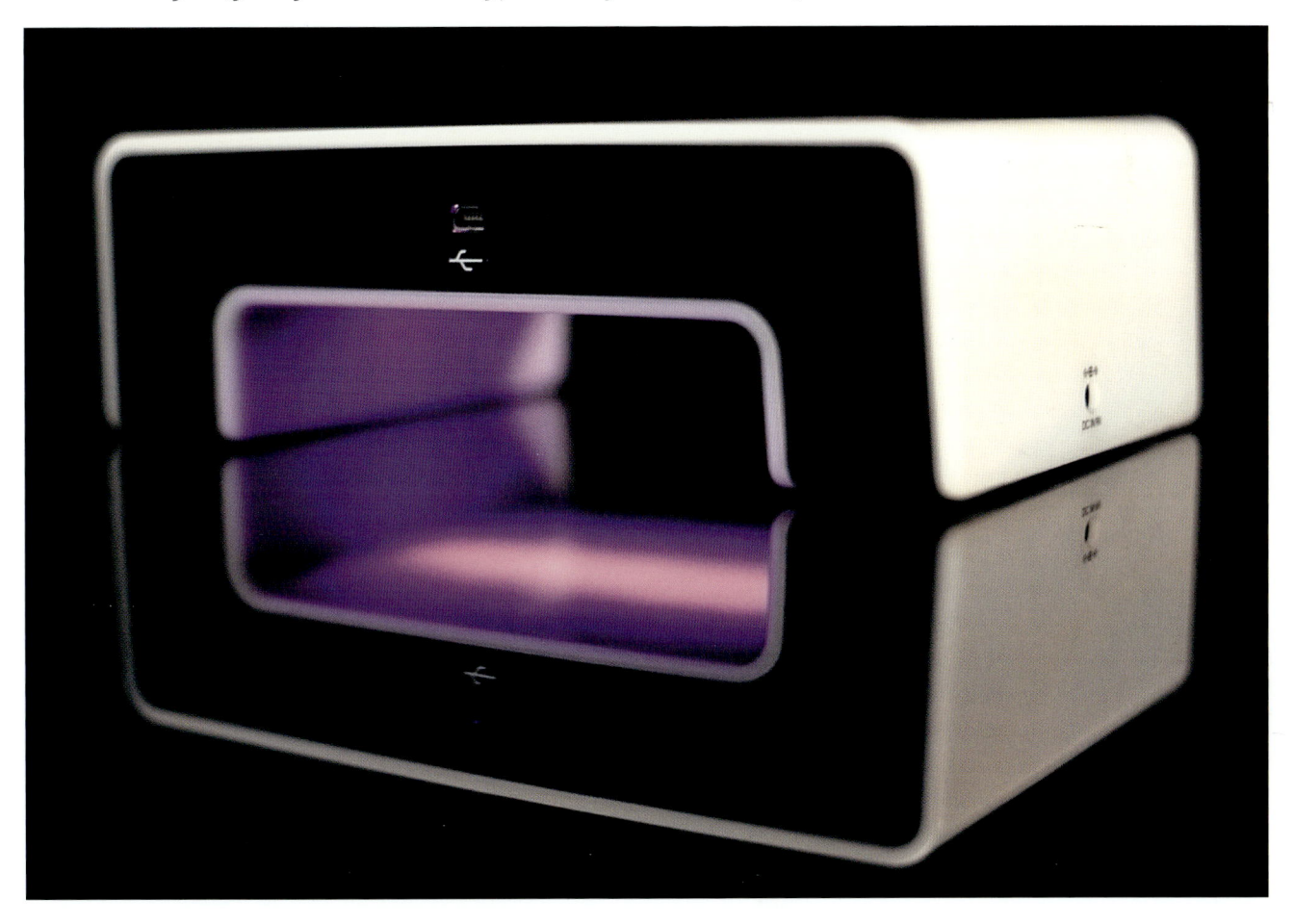

156 LOC8TOR

TECHNO www.loc8tor.com

The Loc8tor has all the properties of a metal magpie, tracking down valuables as much as 600 feet away. Simply attach the small transmitter tag to items that have an annoying habit of going AWOL and the Loc8tor will lead you straight to them. If you lose the Loc8tor itself, well, you've only got yourself to blame.

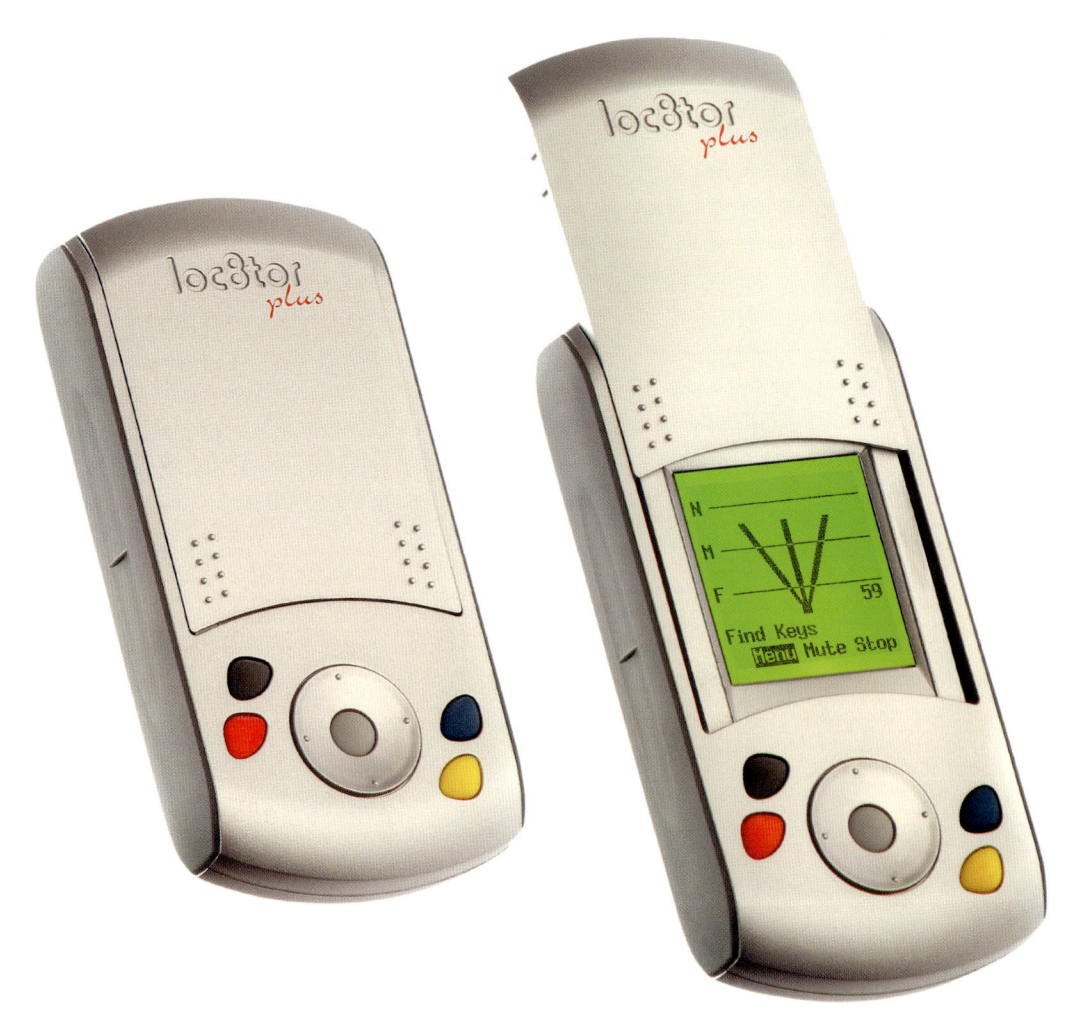

WHY CRY BABY ANALYSER

www.thinkgeek.com www.amazon.com

TECHNO

Any new parent will be familiar with an inconsolable baby. The Why Cry Baby Analyser uses advanced frequency analysis technology to determine why Baby is crying. The verdict appears in less than 20 seconds on the LCD screen. Light and ergonomic, the device comes with a guide to symptoms and a few hints for stimulating Baby's development.

158 REAR VIEW MIRROR CAR CAMERA RECORDER

TECHNO · www.brickhousesecurity.com

The speed and shock of a road incident can leave you disorientated. It's not always easy to think clearly about what you've just experienced, or about who might be to blame. The Rear View Mirror Car Camera Recorder makes sure you don't miss the important details of any bump or shunt. Five cameras record everything that moves whilst the engine is switched on.

ROVIO WOWWEE ROBOTIC HOME SENTRY

159

www.hammacher.com   www.wowwee.com   TECHNO

It's one ugly looking mutt, but it's also the best guard dog you're ever likely to own — the Rovio WowWee Robotic Home Sentry can detect the presence of another human being in your house. Operated by long-distance remote control via the Internet, the bot comes armed with 360-degree vision. It even emails you high resolution images and video footage of its patrols.

HOME

7

162 COME IN & GO AWAY DOORMAT

HOME | www.uncommongoods.com | www.suck.uk.com

The doormat with two faces – one welcoming, the other less so. Made of natural fibre, the Come In & Go Away Doormat lives up to its name, cleverly making use of the same script to give a different message depending on which side you read it from. Sure to get the conversation started on the doorstep.

HIS & HERS KEY HOLDERS

www.uncommongoods.com   www.j-me.co.uk   163   HOME

Who's the one in your household that turns the place upside down looking for the keys? These His & Hers Key Holders should put a stop to the chaos. It's a simple concept (you put your keys in the appropriate figure), elegantly designed by Jaime and Mark Antoniades. An ideal gift for newlyweds and oldlyweds alike.

164 DART COAT HOOKS

HOME

www.perpetualkid.com

www.urbanjunkie.co.uk

When is a dart not a dart? When it's a coat hook. Created by designer Anthony Chrisp, these stainless steel darts can be fixed to the wall to hold jackets, handbags and other accessories. Wall plugs supplied with the kit ensure the darts are easy to affix. Just don't get confused the next time you're in the local pub — the dartboard isn't a cloakroom.

MOBILES BY SALTYANDSWEET

www.etsy.com

HOME

With pin-up women, 'Star Wars' characters, scenes from a riot, zombies and pirates taking on ninjas, these SaltyandSweet mobiles will appeal to big kids like yourself as much as the baby in its crib. Each mobile is elegantly hand cut, creating a dancing cloud of attractive black shapes.

166 EYE CLOCK

HOME | www.hayneedle.com | www.vita-interiors.com

Originally designed in 1957 by George Nelson, and reproduced here by the Vitra Design Museum, the oversized Eye Clock wouldn't look out of place on the set of 'Mad Men'. The clock breaks the shape of the human eye down into a composition of geometric forms. Can be hung on the wall vertically or horizontally.

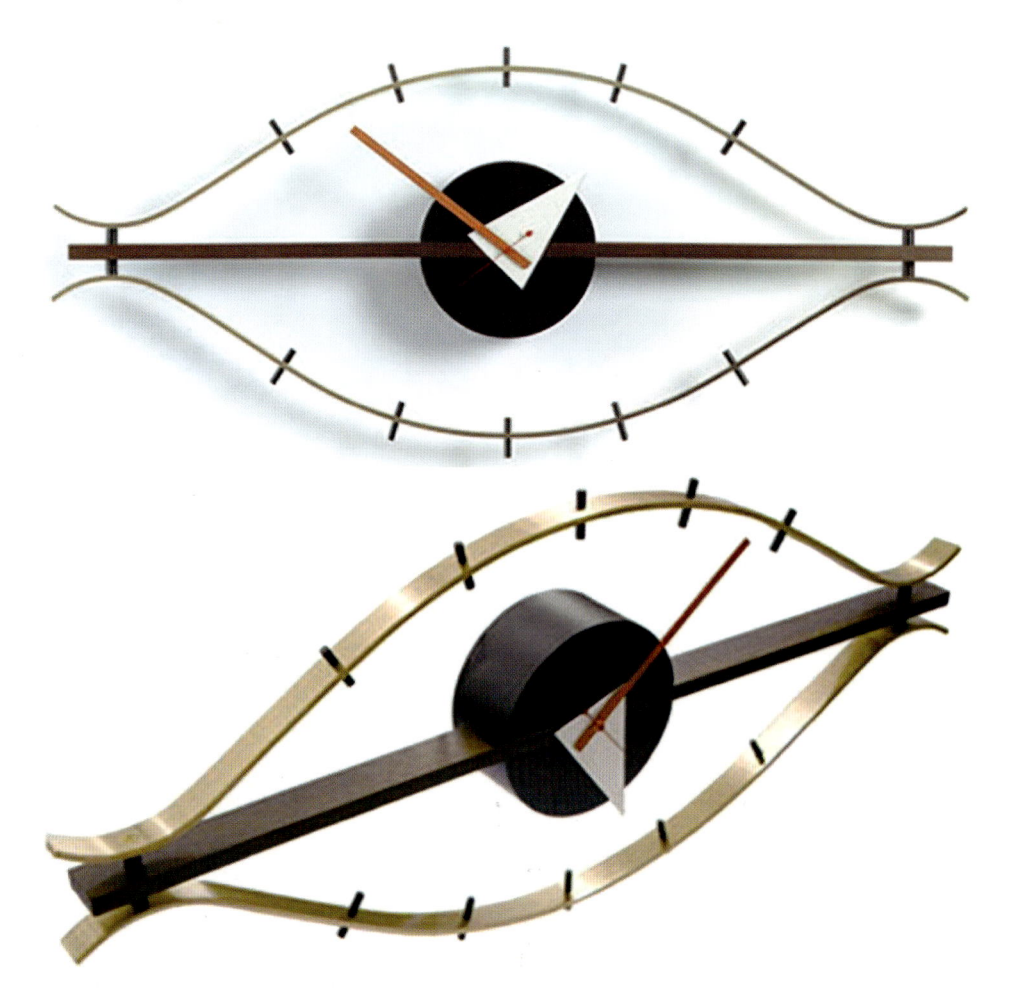

MEMORY GAME ALARM CLOCK

167

www.urbantrendhk.com www.geschenkidee.ch

HOME

Getting up in the morning isn't the easiest thing to do. Even if the body is willing, the brain may refuse to switch on. The Memory Game Alarm Clock should help get you going. The only way to stop the sound is to repeat the colour sequences that light up with the alarm in the right order. It's reminiscent of the 1970s 'Memory Game'. Removing the batteries isn't allowed!

168 LED JELLYFISH MOOD LAMP

HOME

www.thinkgeek.com

www.amazon.co.uk

Nothing chills you out like watching a tank full of jellyfish. But aquariums are a pain to clean and expensive, particularly if you work late and are rarely home. The LED Jellyfish Mood Lamp gives you all the tranquillity of the deep without the stress of the upkeep. The three silicone jellyfish move around the luminous aquarium while the light subtly changes shade. Now just relax.

TEDDY BEAR LAMP

www.gnr8.biz www.suck.uk.com

169

HOME

Poor old Teddy; butchered like some bizarre furry Frankenstein freak. Where once there was a cuddly cutesy head, now we have a light bulb and a dark shade (light and dark – seems to sum it all up rather well). Still, it looks good, and let's be honest, Teddy was getting on a bit; the end was probably nigh anyway...

SLIDE LIGHT

www.gnr8.biz www.suck.uk.com

Snapping family and friends on an old slide camera is great, but so many of those photos end up buried in a drawer, never to see the light of day. The Slide Light provides a brilliant means of showing them off. The backlit, wall-mounted light, made of white-enamelled steel, can be loaded either horizontally or vertically with standard film slides.

INFLATABLE GILDED FRAME

www.brooklyn5and10.com   www.curiosite.com   HOME

OK, so you're not Van Gogh, but you've had a go; that's the main thing. And now, to make your home-painted masterpiece really stand out, put it in an Inflatable Gilded Frame. Whatever it's framing, the blow-up surround brings a breath of modernity (even with its old masters styling), as well as a little light heartedness, to the subject.

UMBRA INVISIBLE BOOKSHELF

HOME — www.amazon.com — www.amazon.co.uk

This bookshelf breaks the normal rules of storage (and, at first glance, the laws of gravity) by appearing to levitate in the air. Simply fix the bracket to the wall, insert the cover of a book and then build the other books up on top. Aside from mesmerising dinner guests with its gravity-defying stunt, the bookshelf works really well on a purely aesthetic level.

JAMES THE BOOKEND

www.lumens.com www.black-blum.com

"Oh James, hold me in your arms, please hold me tight!" shriek the old-fashioned airport novels when hunky Jim saunters into the room. Practical as well as good looking (the ideal man, surely), James is in fact a bookend. Stable, solid – is there any end to his qualities – James blends with any interior thanks to a range of colours including red, orange, black and muted lime.

174 ROCKING SQUARES

HOME | www.roije.com

One glance at the Rocking Square suggests that Dutch designer Frederik Roijé has succeeded in his aim of reinventing traditional furniture. It's a thoroughly modern take on an old favourite: the rocking horse. Made from wood with a durable finish, it comes in three sizes, the largest at 2.5 metres long will happily gallop along with a grown-up rider astride.

SILVERFISH AQUARIUM

175

www.ballertoys.com www.octopusstudios.com HOME

Blessed is the fish that finds itself swimming through the watery excitement of the Silverfish Aquarium — it's worthy of a Bond villain's lair. Six interlinked spheres give your wet pets ample opportunity for cruising. Best of all, the creative flair of the design makes the Silverfish Aquarium a fine, stylish addition to the room.

SNAKE OUT SLIDE CURVED LED BENCH

www.voltexdesign.com

Anything but 'part of the furniture', the Snake Out Slide puts traditional seating designs – both garden and indoor – in stark, right-angled perspective. The single luminous benches measure 123cm long by 43cm wide, and can be pieced together to make much longer seats. Once you've got it where you want it, dim all other lights and the curved bench becomes the centre of attention.

FUNKEY STOOL

www.alexgarnett.com

HOME

When friends arrive or relatives descend, seating 'issues' can afflict even the best prepared host. Somehow the old beanbag serves as a rather sorry stopgap. Wouldn't the Funkey Stool be so much better (and cooler)? The giant computer key combines comfort with 21st century style. Choose your key, 'Pause', 'Esc', 'Home', 'Delete' and more. Line them up and pretend you're really small.

MAXLOUNGE INFLATABLE LOUNGE CHAIR

HOME | www.gadgets.co.uk | www.flashwear.com

Turn on, tune in and flop out: the Maxlounge Inflatable Lounge Chair has got the situation under control. It's an inflatable armchair with two built-in speakers and a jack that will connect you up to your sound system. It breathes much-needed new life into the old phrase 'musical chairs', and even comes with a puncture repair kit, just in case…

WAVE HAMMOCK

www.wave.st www.encompassco.com

Forget coconut trees, the Wave Hammock supports itself. In fact, it looks more like it's floating in space. Constructed from polished stainless steel, the structure filters 90% of the sun's rays, can rotate 360° and comfortably holds two people. The aquamarine sail will make you think the South Seas are lapping gently at your garden – magical!

180 SUSHI PILLOWS

www.thinkgeek.com

www.curiosite.com

So, it's a cushion made from raw fish? No, you've completely misunderstood. It's a pillow that looks like sushi but is in fact... a pillow. Take your pick from salmon, maki, shrimp nigiri or even a giant green edamame, all of them decadently comfy and perfect for the sofa. Warning: these pillows may make you hungry.

SPACE INVADERS WALL STICKERS

www.thinkgeek.com

www.bouf.com

HOME

Remember the early days of video games? Those very basic space invaders chomping their way down the screen. These giant wall stickers will bring it all flooding back. They affix to any smooth flat surface, such as a wall, floor or even chair. Each set contains 13 aliens in four different colours, to be mixed and matched as you see fit.

182	**BLOOD BATH SHOWER CURTAIN**	
HOME	www.perpetualkid.com	www.paramountzone.com

Recreate all the homely comfort (not!) of a Bates Motel en-suite with the Blood Bath Shower Curtain. The bloody hand and footprints bring back the atmosphere of Hitchcock's iconic 1960s movie, but without any of the actual gore. We just hope it doesn't match the rest of your décor.

DROP COAT RACK

183

www.pulpoproducts.com www.thedesigntown.com HOME

We guarantee that the first time you go to hang a coat on the Drop Coat Rack, you'll stop involuntarily, reach for the wall and dab with your finger. Is that wet paint? Your mind says 'don't be ridiculous' but still you reach out. Sleekly shaped from steel, the rack can bear the weight of several coats. It comes in two sizes and seven colours.

184 LED SHOWER LIGHT

HOME | www.amazon.com | www.iwantoneofthose.com

Sticking your hand under the shower to test if the water is the right temperature is so primitive. This is the 21st century. You should be using the LED Shower Light. Combining fun with practicality, the showerhead delivers a stream of light that changes with the temperature of the water. As you might expect, blue means cold and red means hot.

BRIGHTHANDLE LED LIT DOOR HANDLE

www.brighthandle.com · www.ccbrass.com

HOME

No one needs anxiety in the bathroom. It should be a place for relaxation, for reflection even. In particular, you don't need to be stressing about whether or not the door is properly locked. The Brighthandle LED Lit Door Handle removes any doubt: it turns red when the room is occupied; and goes green when the room becomes unoccupied. Works off either the mains or batteries.

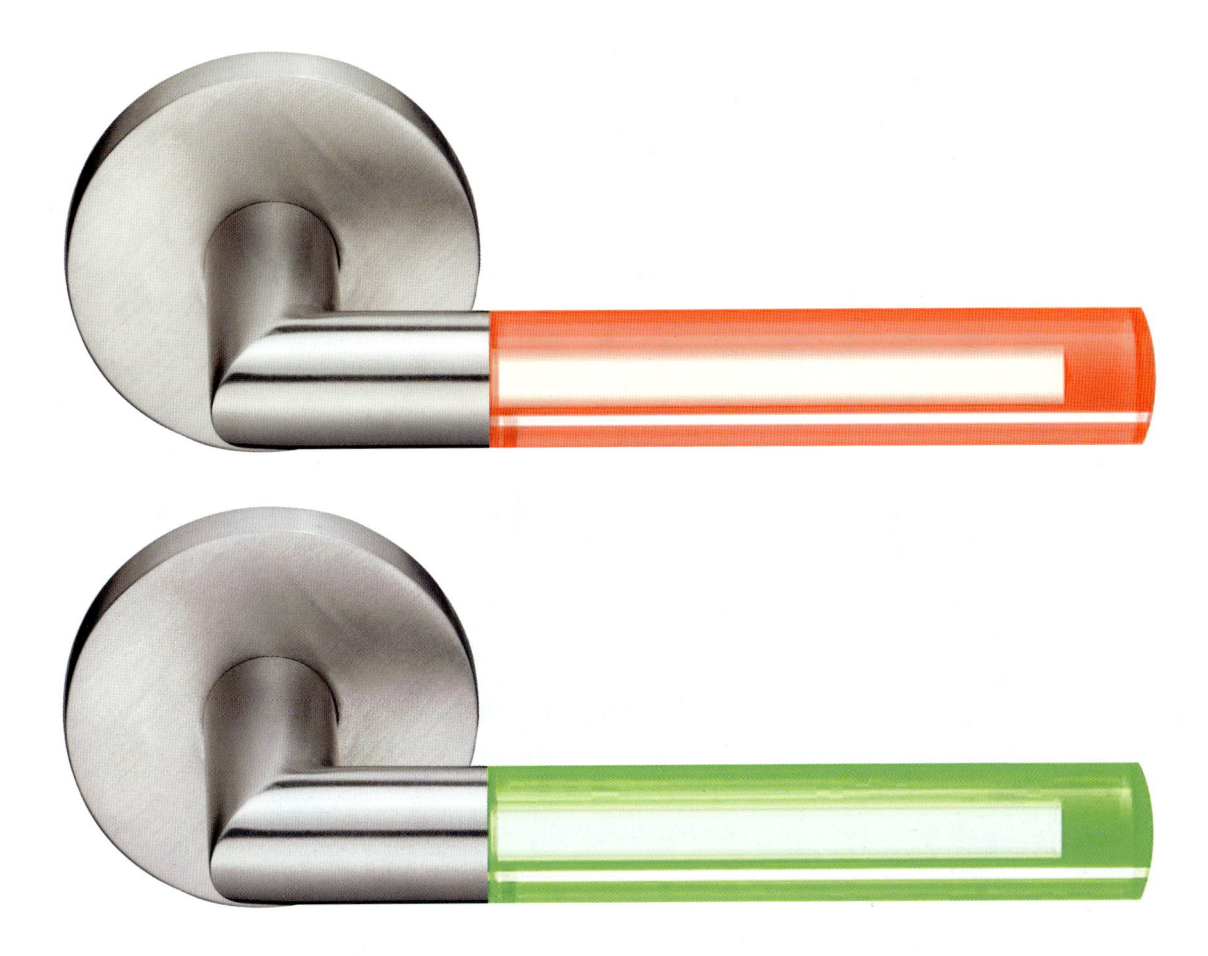

186 GLOW IN THE DARK TOILET ROLL

HOME | www.iwantoneofthose.com | www.glow.co.uk

No one wants to be fumbling around in the dark for toilet rolls (well no one we know anyway), but sometimes it just happens, and for the most inexplicable of reasons. Just in case, you really should have some phosphorescent toilet rolls on standby. Aside from acting like a paper beacon in the dead of night, they also make potty training more fun for the little ones.

LAVNAV TOILET NIGHTLIGHT

www.arkon.com   www.gadgets.co.uk

As any woman knows, men are hopeless at firing into an open toilet, particularly when the seat is down... and it's dark. And, as any man knows, aiming straight isn't all that easy. The Lavnav Toilet seat is fitted with a movement sensor that activates a green light when the seat is down, and a red light when it is up, it also shines a light on the bottom of the bowl giving gents a target to aim for.

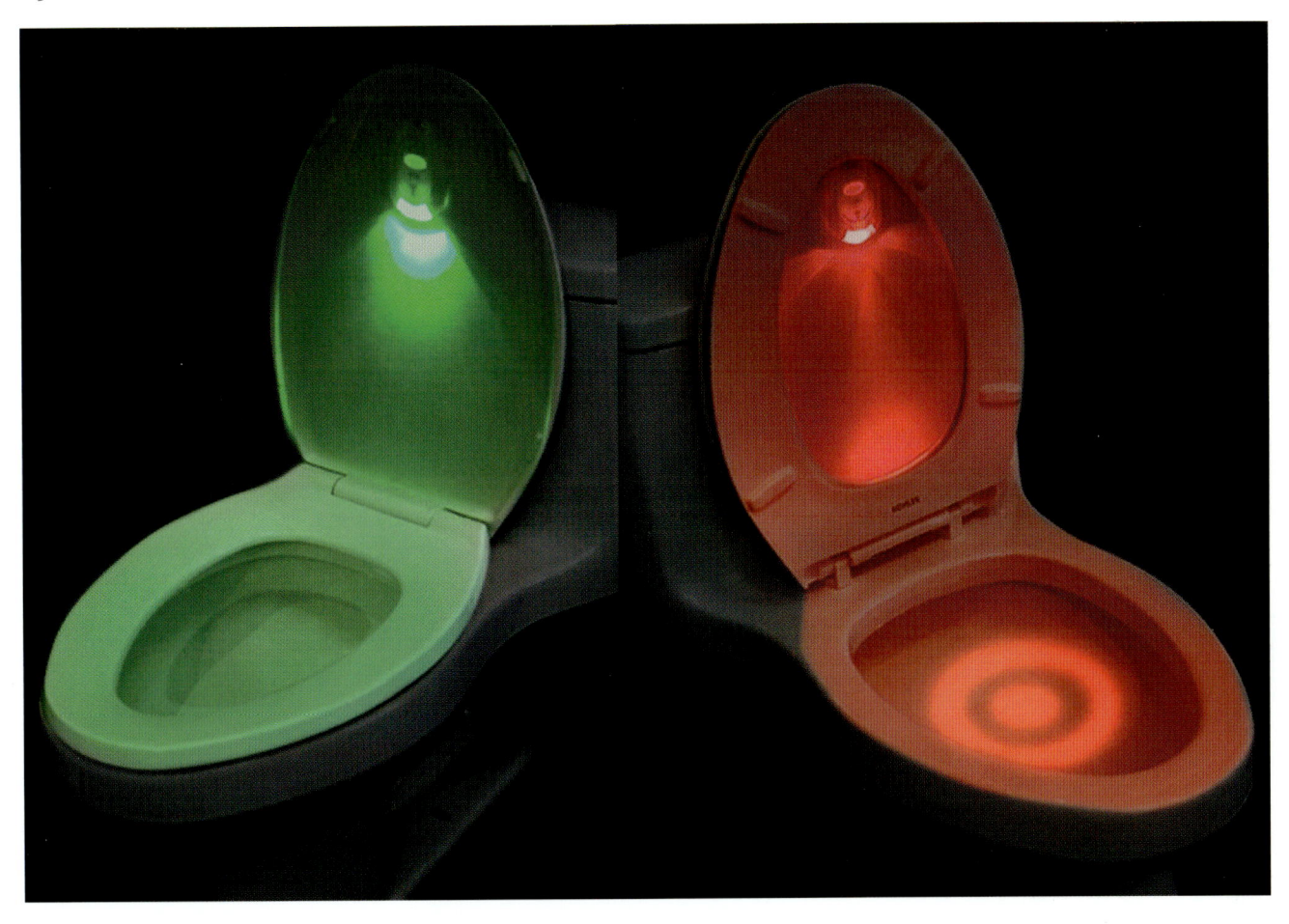

188 PET GOLDFISH BIN BAGS

HOME | www.shinyshack.com | www.suck.uk.com

Pet Goldfish Bin Bags bring a little light relief to the routine chores of bin collection day. Biodegradable and wholly free from toxic materials, the bags have a capacity of 100 litres and are sold in packs of 12. Just don't be surprised to find an audience of neighbourhood cats sitting in front of your bins, tentatively pawing at the oversized fish.

STAR WARS R2-D2 TRASHCAN

www.forbiddenplanet.co.uk

Why hasn't someone thought of this before? R2-D2 always did look like the most impractical robot ever designed – steps and curbs were a complete no-go. His re-imagining as a pedal bin seems to make perfect sense. At 60cm high, the limited edition bin is a faithful replica of the original 'Star Wars' character.

INDEX

INDEX BY PURPOSE

Do you want to thank the Joneses for that fantastic dinner? Or congratulate your colleague on being promoted? Here you'll find a mine of gift ideas for all those important occasions.

IMPRESSING: 11, 13, 14, 24, 26, 28, 30, 32, 33, 35, 39, 43, 45, 50, 52, 62, 63, 66, 67, 77, 80, 90, 91, 92, 93, 94, 95, 99, 100, 101, 102, 103, 106, 121, 129, 131, 132, 133, 136, 137, 139, 141, 144, 145, 146, 147, 149, 154, 156, 158, 166, 169, 170, 171, 174, 175, 176, 181, 185, 189

KITTING OUT: 10, 11, 12, 13, 14, 15, 17, 18, 19, 20, 21, 22, 25, 30, 31, 32, 33, 34, 71, 72, 73, 75, 79, 81, 83, 100, 114, 116, 118, 119, 124, 137, 140, 141, 142, 143, 145, 146, 152, 153, 154, 159, 163, 164, 165, 166, 172, 173, 177, 178, 183, 184

WELLBEING: 38, 42, 44, 48, 49, 50, 51, 56, 96, 98, 120, 127, 131, 143, 150, 151, 155, 157, 168, 179

HUMOUR: 10, 12, 15, 16, 17, 22, 23, 24, 25, 26, 27, 29, 31, 33, 35, 38, 40, 41, 47, 52, 53, 60, 61, 66, 68, 69, 70, 73, 74, 75, 76, 78, 80, 82, 86, 87, 88, 89, 96, 97, 104, 108, 109, 110, 111, 115, 117, 118, 121, 122, 123, 125, 128, 130, 138, 142, 148, 152, 162, 164, 165, 180, 182, 186, 187, 188

TAKING THE MICKEY: 21, 23, 27, 29, 68, 69, 78, 82, 83, 88, 89, 104, 109, 110, 121, 130, 148, 157, 165, 167, 187

SHOWING AFFECTION: 18, 19, 46, 54, 55, 57, 74, 76, 77, 120, 126, 127, 128, 129, 140, 169

INDEX BY TYPE OF RECIPIENT

A green chum? A glamorous mother-in-law who loves to party? A geeky neighbour? You can be sure of delighting the recipient by finding the perfect present for that particular person.

LOVERS: 52, 54, 55, 56, 57, 74, 76, 77, 126, 127, 129, 130, 131

ANXIOUS: 13, 26, 42, 47, 49, 50, 51, 66, 81, 96, 115, 156, 157, 159, 163, 168, 182, 186

PET: 82, 83, 108, 109, 110, 111, 175

ADVENTUROUS: 13, 26, 71, 79, 98, 99, 118, 128, 132, 133, 139, 154

YOUNG KIDS: 15, 16, 17, 18, 19, 80, 81, 86, 87, 89, 90, 91, 94, 95, 102, 103, 105, 151, 165, 167, 169

CREATIVE: 10, 15, 16, 18, 27, 28, 35, 40, 44, 73, 80, 86, 90, 114, 117, 162, 164, 171, 181, 188

GREEN: 20, 65, 119, 125, 129, 155, 175

FASHION VICTIM: 28, 38, 43, 53, 62, 63, 64, 67, 70

PARTY ANIMAL: 12, 21, 24, 29, 32, 39, 62, 63, 68, 69, 74, 75, 76, 88, 131, 152, 153

GEEK: 19, 27, 41, 60, 67, 95, 100, 103, 114, 116, 119, 121, 122, 123, 124, 125, 132, 136, 137, 138, 140, 141, 143, 145, 146, 148, 149, 154, 156, 157, 159, 177, 180, 184, 185, 187, 189

BIG KID: 17, 29, 31, 33, 64, 78, 87, 90, 91, 92, 93, 94, 95, 97, 102, 103, 104, 105, 121, 122, 138, 150, 151, 164, 165, 167, 168, 169, 181, 186

DESIGN FREAK: 11, 14, 19, 25, 30, 31, 33, 35, 92, 93, 101, 133, 146, 165, 166, 170, 172, 174, 176, 177, 183

MUSIC ADDICT: 75, 142, 143, 144, 145, 147, 153

NOSTALGIA-LOVER: 45, 46, 61, 65, 120, 122, 130, 136, 140, 152, 169, 170, 181, 189

SPORTY: 14, 48, 49, 78, 96, 98, 99, 100, 101, 106, 107, 118

INDEX BY OCCASION

Christmas, Valentine's Day, birthday, new baby... a gift for every occasion.

HOUSE-WARMING: 10, 11, 12, 14, 18, 20, 21, 22, 25, 29, 30, 32, 33, 35, 88, 119, 150, 155, 162, 163, 164, 167, 171, 172, 173, 177, 178, 180, 181, 182, 183, 184, 185, 186, 188, 189

RETIREMENT: 24, 79, 92, 96, 102, 104, 118, 139, 143, 152, 154, 158

MOTHER'S DAY: 30, 31, 34, 42, 43, 44, 51, 60, 63, 65, 120, 140

FATHER'S DAY: 13, 14, 23, 24, 32, 38, 39, 49, 51, 66, 67, 68, 69, 70, 80, 91, 97, 117, 137, 141, 156

WEDDING: 10, 11, 18, 25, 30, 33, 35, 50, 76, 77, 121, 131, 170, 174, 175, 176, 179, 184

NEW BABY: 15, 19, 46, 47, 81, 86, 87, 89, 151, 157, 165, 169

VALENTINE'S DAY: 38, 45, 52, 53, 55, 57, 62, 63, 65, 74, 76, 77, 78, 121, 126, 127, 131

INDEX BY PRICE

Have you set yourself a firm budget? No problem: here, you'll find a host of possibilities to suit your pocket.

BUDGET: 12, 15, 16, 17, 20, 21, 22, 23, 24, 26, 26, 27, 28, 28, 29, 31, 32, 33, 35, 38, 39, 40, 41, 42, 43, 44, 46, 48, 51, 52, 54, 55, 56, 57, 60, 61, 64, 65, 68, 69, 70, 74, 75, 77, 78, 79, 81, 82, 83, 86, 87, 88, 93, 94, 95, 96, 102, 104, 105, 108, 109, 110, 111, 114, 115, 116, 117, 119, 120, 122, 123, 124, 125, 126, 127, 128, 129, 130, 131, 132, 138, 140, 145, 151, 154, 157, 162, 163, 164, 165, 168, 171, 172, 173, 178, 180, 181, 182, 184, 186, 187, 188

AFFORDABLE: 10, 14, 18, 19, 25, 30, 34, 45, 47, 49, 50, 53, 53, 62, 63, 66, 71, 72, 73, 80, 89, 100, 101, 103, 106, 107, 118, 121, 136, 137, 139, 142, 144, 148, 152, 153, 155, 156, 166, 167, 169, 183, 185, 189

EXTRAVAGANT: 11, 13, 67, 76, 90, 91, 92, 97, 98, 99, 133, 141, 143, 146, 147, 149, 150, 158, 159, 170, 174, 175, 176, 177, 179

PICTURE CREDITS

10: ©Fred & Friends; 11: ©Thout; 12: ©Fred & Friends; 13: ©Wenger SA; 14: © Alex Garnett; 15: ©Fred & Friends; 16: ©Firebox.com; 17: ©Modern-twist, Inc; 18: ©LC Premiums, Ltd; 19: ©iiamo; 20: ©Firebox.com; 21: ©I Want One Of Those, Ltd; 22: ©Think Geek, Inc; 23: ©Fred & Friends; 24: ©Firebox.com; 25: ©David-Louis Hendley; 26: ©Edible; 27: ©Thinkofthe; 28: ©LaboGroup; 29: ©Tadenberg; 30: ©Eva Solo; 31: ©Suck UK, Ltd; 32: ©I Want One Of Those, Ltd; 33: ©Liz Goulet Dubois; 34: ©Firebox.com; 35: ©Urban Trend, Ltd; 38: ©ThinkGeek, Inc; 39: ©The Beer Soap Company; 40: ©Dxion; 41: ©Digital Soaps; 42: ©Gelicity UK Ltd; 43: ©Orange Planet BV; 44: ©Firebox.com; 45: ©Scents of Time; 46: ©The Baby Art Company; 47: ©Art de toilette; 48: ©Kinesio Holding Corporation; 49: ©Clinically Fit Inc.; 50: ©Ixit SAS Tiao; 51: ©Brookstone, Inc; 52: ©Firebox.com; 53: ©Kiskin; 54: ©Condoms.co.uk; 55: ©Atypyk; 56: ©Atypyk; 57: ©The Kama Sutra Company; 60: ©ThinkGeek, Inc; 61: ©ThinkGeek, Inc; 62: ©LumiGram SARL; 63: ©Beau & Bien; 64: ©Rodier / Attakus America Inc.; 65: ©Anna Built; 66: ©Sruli Recht; 67: ©Charles & Marie KG ; 68: ©Holster Up Inc; 69: ©Underdevelopment Inc; 70: ©Suck UK Ltd; 71: ©Rodrigo Alonso Schramm; 72: ©Senz Umbrellas BV; 73: ©Art. Lebedev Studio; 74: ©ThinkGeek, Inc; 75: ©ThinkGeek, Inc; 76: ©Tateossian Ltd; 77: ©Radius Design; 78: ©Cultstyles GmbH; 79: ©Sun-Mate Corp; 80: ©Aamu Song & Johan Olin ; 81: ©Kelly Forsyth-Gibson; 82: ©BuyCostumes.com; 83: ©Doggles, LLC; 86: ©Natural Products (UK) Ltd; 87: ©Compost Communications Ltd; 88: ©Anycostume.co.uk; 89: ©Julie Arma; 90: ©ÜberStix LLC; 91: ©Mightymast Leisure ; 92: ©Straight Up Chess; 93: ©MegaHouse Corporation; 94: ©Elenco Electronics Inc; 95: ©Radica; 96: ©Gadget Epoint Ltd; 97: ©Funtrack Ltd; 98: ©RolerBall.co.uk; 99: ©Phoenix-Fly design; 100: ©MonkeyLectric; 101: ©UK Innovations GP LTD; 102: ©Ravensburger; 103: ©Fascinations; 104: ©ThinkGeek, Inc; 105: ©WowWee Group Ltd; 106: ©SBI Enterprises; 107: ©Freeline Sports, Inc. ; 108: ©ThinkGeek, Inc; 109: ©Suck UK Ltd; 110: ©Sicandra Co; 111: ©R2 Solutions; 114: ©Suck UK Ltd; 115: ©Fred & Friends; 116: ©Suck UK Ltd; 117: ©Suck UK Ltd; 118: ©Nauticalia Ltd; 119: ©ThinkGeek, Inc; 120: ©Inclusive Technology Ltd; 121: ©Maker Media Division, O'Reilly Media, Inc; 122: ©Mimoco, Inc; 123: ©I Want One Of Those Ltd; 124: ©Brando Workshop; 125: ©Find Me A Gift; 126: ©Atypyk; 127: ©I Want One Of Those Ltd; 128: ©details, produkte + ideen; 129: ©Suck UK Ltd; 130: ©Curiosite.com; 131: ©Firebox.com; 132: ©Urban Mapping Inc.; 133: ©DotDrops; 136: ©Yoostar Entertainment Group, Inc; 137: ©Promotion and Display Technology Ltd; 138: ©Firebox.com; 139: ©Firebox.com; 140: ©I Want One Of Those Ltd; 141: ©Vuzix Corporation; 142: ©D.R; 143: ©Sony Corporation; 144: ©D.R; 145: ©Fred & Friends; 146: ©NovelQuest Enterprises Inc.; 147: ©Beamz Interactive, Inc; 148: ©KeyboardForBlondes.com; 149: ©Logitech; 150: ©Sega Toys; 151: ©I Want One Of Those Ltd; 152: ©Firebox.com; 153: ©Firebox.com; 154: ©Boston Ideas, LLC; 155: ©Toby Summerskill; 156: ©Loc8tor Ltd; 157: ©ThinkGeek, Inc; 158: ©BrickHouseSecurity.com; 159: ©Hammacher Schlemmer & Company; 162: ©Suck UK Ltd; 163: ©j-me; 164: ©Suck UK Ltd; 165: ©SaltyandSweet; 166: ©George Nelson for Vitra; 167: ©Geschenkidee.ch GmbH ; 168: ©ThinkGeek, Inc; 169: ©Suck UK Ltd; 170: ©Suck UK Ltd; 171: ©Curiosite.com; 172: ©Umbra Ltd; 173: ©Eureka SARL; 174: ©Frederik Roijé; 175: ©Octopus Studio SARL; 176: ©Voltex; 177: ©Alex Garnett; 178: ©Maxlounge.net; 179: ©Eric Nyberg & Gustav Storm for Royal Botania; 180: ©Curiosite.com; 181: ©Curiosite.com; 182: ©Paramountzone.com; 183: ©Pulpo; 184: ©I Want One Of Those Ltd; 185: ©Emtek; 186: ©I Want One Of Those Ltd; 187: ©I Want One Of Those Ltd; 188: ©Suck UK Ltd; 189: ©Forbidden Planet International.

Chantal Allès is a journalist. The press, radio, television… her curiosity has taken her into numerous domains. A Reporter before becoming Editor of several television programmes on culture and society, she is particularly interested in trends in contemporary society, with a predilection for the idiosyncratic and the absurd.

Anne Kerloc'h is a journalist, columnist and author of several books: documentaries, coffee-table books and young fiction. Chief columnist on the daily "20 Minutes", she has also written for Bayard Presse, Charlie-Hebdo and Futur(e)s, and runs writing workshops for young people and adults. With a taste for people, words, humour and the absurdities of life.